T0139061

GeoSensor Networks

GeoSensor Networks

EDITED BY
Anthony Stefanidis
Silvia Nittel

CRC PRESS

Boca Raton London New York Washington, D.C.

Library of Congress Cataloging-in-Publication Data

Catalog record is available from the Library of Congress

Direct all inquiries to CRC Press LLC, 2000 N.W. Corporate Blvd., Boca Raton, Florida 33431.

Visit the CRC Press Web site at www.crcpress.com

International Standard Book Number 0-41532-404-1
Library of Congress Card Number
Printed in the United States of America 1 2 3 4 5 6 7 8 9 0
Printed on acid-free paper

Preface

Advances in sensor technology and deployment strategies are revolutionizing the way that geospatial information is collected and analyzed. For example, cameras and GPS sensors on-board static or mobile platforms have the ability to provide continuous streams of geospatially-rich information. Furthermore, with the advent of nano-technology it becomes feasible and economically viable to develop and deploy low-cost, low-power devices that are general-purpose computing platforms with multi-purpose on-board sensing and wireless communications capabilities. All these types of sensors may act collaboratively as nodes within broader network configurations. Such configurations may range in scale from few cameras monitoring traffic to thousands of nodes monitoring an ecosystem.

When drafting the call for papers that resulted in this book we left on purpose the term “geosensor network” somewhat undefined, as we wanted it to be determined by the research communities that are providing the pieces of its puzzle. However, despite this lack of a formal definition there is a very clear inherent understanding of what a geosensor network is. In geosensor networks the geospatial content of the information collected by a sensor network is fundamental for the analysis of feeds. Thus, a geosensor network may be loosely defined as a sensor network that monitors phenomena in a geographic space. This space may range in scale from the confined environment of a room to the highly complex dynamics of a large ecosystem.

With this emerging sensor deployment reality we are faced with substantial research challenges related to the collection, management, analysis, and delivery of real-time geospatial information using distributed geosensor networks. This book offers a collection of papers that address some of these issues. The papers included here were presented at the first GeoSensor Networks workshop, held in Portland, Maine, in October 2003. This fully refereed workshop brought together thirty-two researchers from diverse research domains, including spatial databases and spatial information modeling, robotics and digital image analysis, mobile computing, operating systems, database management, and environmental applications.

Our objective was to provide a forum for experts from these overlapping communities to exchange ideas and share knowledge, and we hope that this book will showcase this spirit of collaboration. The papers in this volume have been grouped in four categories, reflecting major aspects of geosensor networks, namely databases, image processing, computer networks, and some application examples. Combined, these papers offer an excellent snapshot of the state-of-the-art in these areas, and support a fair evaluation of the current capabilities and emerging challenges for geosensor networks. Additional information on our workshop, including program and presentation material, may be found at the corresponding web site www.spatial.maine.edu/~gsn03/.

We would like to thank the authors of the papers included in this volume, and additional invited presenters at the workshop for their valuable contributions, the program committee members for their input, and Working Group V/5 of the International Society for Photogrammetry and Remote Sensing. We would also like to thank our students and colleagues at the University of Maine who assisted in the organization of the workshop in many ways, and especially Mr. Charalampos Georgiadis who assisted in the preparation of this volume. We would like to particularly acknowledge the help and guidance of Professor Max Egenhofer, who was instrumental in the realization of this event. Lastly, we would like to acknowledge the National Science Foundation for supporting the workshop through grant EIA-9876707.

February 2004 Anthony Stefanidis and Silvia Nittel

Workshop Organization

Co-Organizers:
Silvia Nittel and *Anthony Stefanidis*
National Center for Geographic Information and Analysis
University of Maine

Workshop Steering Committee:
Chaitan Baru, San Diego Supercomputer Center
Deborah Estrin, University of California, Los Angeles
Mike Franklin, University of California, Berkeley
Johannes Gehrke, Cornell University
Mike Goodchild, University of California, Santa Barbara
Nick Koudas, AT&T Research
Richard Muntz, University of California, Los Angeles
Silvia Nittel, University of Maine (Workshop co-chair)
Anthony Stefanidis, University of Maine (Workshop co-chair)
Seth Teller, Massachusetts Institute of Technology
Mubarak Shah, University of Central Florida

Editors

Anthony Stefanidis is Assistant Professor in the Department of Spatial Information Science and Engineering, and the National Center for Geographic Information and Analysis (NCGIA) at the University of Maine. He holds a Dipl. Eng. degree from the National Technical University of Athens, Greece, and M.S. and Ph.D. degrees from The Ohio State University. Before joining the University of Maine he spent two years as senior researcher at the Swiss Federal Institute of Technology (ETH) in Zurich. His research activities focus on digital image and video analysis for geospatial applications, including optical sensor networks. His work is currently sponsored by the National Science Foundation and the National Geospatial-Intelligence Agency. In addition to numerous publications, Tony has edited one more book in his area of expertise, and has contributed chapters to several other books.

Silvia Nittel is Assistant Professor with the Department of Spatial Information Science and Engineering, and the National Center for Geographic Information and Analysis (NCGIA) at the University of Maine. She obtained her Ph.D. in Computer Science at the University of Zurich in 1994, and she specialized in non-traditional database system architectures. Silvia spent several years as postdoctoral researcher and later co-director of the UCLA Data Mining Lab. At UCLA, she worked on high performance data mining tools for knowledge extraction from raster satellite data sets, heterogeneous data integration for geoscientific data and interoperability issues. She was the project lead of a large NASA-funded research effort at UCLA. In 2001, Silvia joined the University of Maine, and has since focused on data management for sensor networks, geosensor networks, and mobile computing.

Contents

GeoSensor Networks and Virtual GeoReality

Silvia Nittel and Anthony Stefanidis

Department of Spatial Information Science & Engineering
National Center for Geographic Information and Analysis
Orono, ME 04469-5711
{nittel, tony}@spatial.maine.edu

ABSTRACT

The use of sensor networks is revolutionizing the way that geospatial information is collected and analyzed. The old paradigm of calibrated sensors collecting information in a highly-controlled deployment strategies is now substituted by wireless networks of diverse sensors that collect information feeds that vary substantially in content, resolution, and accuracy. This evolution is bringing forward substantial challenges in terms of data management and analysis, but at the same time introduces up to date unparalleled scene modeling capabilities. In this paper we provide a brief summary of workshop findings, and introduce our vision of the effect that geosensor networks will have on the communication, access, and modeling of geospatial information

1. INTRODUCTION

Advances in sensor technology and deployment strategies are revolutionizing the way that geospatial information is collected and analyzed. For example, cameras and GPS sensors on-board static or mobile platforms have the ability to provide continuous streams of geospatially-rich information. With the advent of nanotechnology it also becomes feasible and economically viable to develop and deploy low-cost, low-power devices that are general-purpose computing platforms with multi-purpose on-board sensing and wireless communications capabilities.

These advances are introducing a novel data collection scheme, with continuous feeds of data from distributed sensors, covering a broader area of interest. This emerging data collection scheme is introducing interesting research challenges related to information integration and the development of infrastructures for systems comprising numerous sensor nodes. These types of sensors may act collaboratively within broader network configurations that range in scale from a few cameras monitoring traffic to thousands of nodes monitoring an ecosystem. The challenge of sensor networks is to aggregate sensor nodes into computational infrastructures that are able to produce globally meaningful information from raw local data obtained by individual sensor nodes. Emerging applications are rather diverse in terms of their focus, ranging for example from the use of sensor feeds for environmental applications [Ailamaki et al., 2003] and wildlife habitat monitoring [Juang et al., 2002; Mainwaring et al., 2002] to vehicle [Pister et al., 2002] and structure monitoring [Lin et al., 2002], and even a kindergarten environment [Chen et al., 2002].

This short paper is meant to provide both a brief summary of findings of this workshop and a vision of the effect that geosensor networks can have on the communication, access, and modeling of geospatial information. In Section 2 we discuss the evolution of geospatial data collection from traditional approaches to geosensor networks. In Sections 3 and 4 we address sensor network programming using DBMS and discuss the issues of scale and mobility in sensor networks, as they were presented in the GSN workshop. In Section 5 we present our vision of geospatial information modeling in Virtual GeoReality, and follow with some concluding remarks in section 6.

2. GEOSENSOR NETWORKS

A geosensor network can be loosely defined as a sensor network that monitors phenomena in geographic space, and in which the geospatial content of the information collected, aggregated, analyzed, and monitored by a sensor network is fundamental. Analysis and aggregation may be performed locally in real-time by the sensor nodes or between sensor nodes, or off-line in several distributed, in-situ or centralized repositories. Regardless of where these processes take place the spatial aspect is dominant in one or both of the following levels:

- *Content level,* as it may be the dominant content of the information collected by the sensors (e.g. sensors recording the movement or deformation of objects), or
- *Analysis level,* as the spatial distribution of sensors may provide the integrative layer to support the analysis of the collected information (e.g. analyzing the spatial distribution of chemical leak feeds to determine the extent and source of a contamination).

The geographic space covered by the sensor network, or analyzed through its measurements, may range in scale from the confined environment of a room to the highly complex dynamics of an ecosystem region.

The use of sensor networks for geospatial applications is not really new. Satellites and aerial cameras have been providing periodic coverage of the earth during the last few decades. However, the evolution of sensing devices [Helerstein et al., 2003] is revolutionizing geospatial applications. The old paradigm of calibrated sensors collecting information in a highly-controlled deployment strategies is now substituted by wireless networks of diverse sensors. This evolution has a profound effect on the nature of collected datasets:

- Homogeneous collections of data (e.g. collections of imagery) are now substituted by heterogeneous feeds for an area of interest (e.g. video *and* temperature feeds).
- Regularly sampled datasets (e.g. coordinates of similar accuracy in a regular grid) are substituted by pieces of information that vary substantially in content, resolution, and accuracy (e.g. feeds from few distinct irregularly distributed locations with sensors of varying accuracy).
- Information becomes increasingly spatiotemporal instead of just spatial, as sensor feeds capture the evolution over time of the properties they monitor.

This evolution is bringing forward substantial challenges in terms of data management and analysis, but at the same time introduces up to date unparalleled scene modeling capabilities.

3. PROGRAMMING SENSOR NETWORKS USING DBMS TECHNOLOGY

It is a common assumption in the database community that programming sensor networks is hard, and database management system (DBMS) technology with its characteristics of declarative data models, query languages and automatic query optimization makes the job of programming sensor networks significantly simpler. DBMS-style query execution over sensor networks is developed with the requirement that queries are formalized in such a way that their execution plans over the sensor network infrastructure are automatically optimizable by the DBMS.

The main optimization criterion is energy-efficient processing of information since batteries are typically not renewed during the lifetime of an application deployment. Since the transmission of data between sensor nodes is costly with regard to energy consumption, optimization attempts to minimize communication between nodes while guaranteeing quality of service. Strategies include minimization of data acquisition, i.e. instructing sensor nodes to only generate (sample) the data that is necessary for a query, or to only forward new values that are within a significant threshold change of the current sampling values. Another strategy is to exploit automatic operator reordering during query processing so that operators that are 'cheaper' (i.e. lower drain on energy to obtain a sensor sample) are evaluated first, and sampling of more 'expensive' sensors for a conjunctive predicate can be avoided. Other strategies are compressing values so that less data is transmitted between nodes, or suppressing values within a temporal coherency tolerance.

Today, power consumption is driven by sampling sensor values, and listening to queries. Minimizing the listing time of sensor nodes allows them to only wake up and synchronize for very short periods of time. With such a massively distributed computing system the notion of synchronized system time is a major challenge. Also, sampling frequency can be adapted over time to prolong the battery lifetime of sensor nodes.

4. SCALE AND MOBILITY OF SENSOR NODES

The scale of sensor data collection and processing is a significant challenge in geosensor networks. Varying scales of sensor data collection and processing are required for different aspects of a problem or even a particular user. The issue matters with regard to sensor node locations and their distribution density, the size of regions of interest, and intervals of sampling. Also, user and application needs play a significant role as such to collect raw data, statistical data, or models, and the level of quality of service such as freshness of data, response time, etc.

To enable multi-resolution queries, different epoch sizes can be assigned to different spatial areas of the network. Shorter epochs enable a higher frequency data sampling and aggregation. Another alternative consists of a group-based routing tree construction. A 'group' is a set of sensors that e.g. exhibits the same capabilities (e.g. temperature sensing), and the routing tree consists of parent-child nodes of the same group while all nodes are collocated. This decreases the number of messages a parent node has to send, and the number of queries to respond to. Simulation results demonstrate that this mechanism works well for a small number of different groups, but a larger number of members per group.

For today's prototypes, the assumption is made that sensor nodes are stationary for the time being. However, it is most likely that sensors are mobile by either being self-propelled or being attached to moving objects. In the environmental domain for example, sensors may be floating in a drainage or be carried by the wind in storms. Network protocols contain built-in mechanisms to construct flexible routing trees despite the mobility of sensor nodes. Nevertheless, sensor nodes need to be able to geolocate their own position with sufficient accuracy, a problem that is still open today. Current research work in robotics with regard to self localization of robots could be leveraged [Howard et al., 2003]. Likely, sensors nodes are rarely located at exactly the position that is necessary for a spatial region query in the geographic space. Mappings between higher-level spatial user predicates and actual physical sensor node locations are of interest, and also constructing an optimal routing tree for a specific spatial query predicate (see the paper by Goldin et al. in this volume). Furthermore, the density of sensor nodes needs to be mapped to different application resolution needs. Dense deployment of sensor nodes is economically not viable. Mechanisms such as robots fixing density problems by 'dropping' sensor nodes in low density areas might be a more flexible and economic solution.

5. SENSOR NETWORKS ENABLING VIRTUAL GEOREALITY

Communicating the content of geospatial databases has evolved from static representations (e.g. maps) to complex virtual reality models. The development of realistic virtual reality (VR) models of urban environments has been the topic of substantial research efforts in the last few years. One of the premier efforts in this direction is the collaborative effort of the groups of Bill Jepson and Richard Muntz at UCLA for the development of *Virtual LA*, a large-scale virtual model of the city of Los Angeles (see e.g. [Jepson et al., 1996] and the web site www.aud.ucla.edu/~bill/UST.html). The photorealistic 3D model of Los Angeles was created using aerial and street-level imagery, and is used to support a variety of cross-disciplinary simulations (e.g. evaluating urban planning, and rehearsing emergency response actions). From a research point of view the major strength of this effort lies in the development of a system to support interactive navigation over the entire model by integrating many smaller models (over a dozen models) into a large virtual environment.

Other notable efforts focus on image analysis issues to create 3D urban scene models. They include the work of [Brenner, 2000] on the automatic 3D reconstruction of complex urban scenes using height data from airborne laser scanning and the groundplans of buildings as they are provided by existing 2D GIS or map data. Height data are used to create a digital elevation model (DEM) of the city, and a photorealistic virtual city model is generated by projecting aerial or terrestrial images onto this DEM. This approach has been used to create a virtual model of the city of Stuttgart (Germany), covering more than 5000 buildings in an area of 2km x 3km [Haala & Brenner, 1999]. Before the work of the Stuttgart group, the group of Gruen at ETH (Zurich) had worked on the integration of terrain imagery and aerial-sensor-derived 3D city models [Gruen & Wang, 1999]. Similar approaches have been followed in the UK to develop virtual models of the city of Bath, covering several square kilometers of the historic center of the city at sub-meter resolution [Day et al., 1996], in Austria to establish models of the cities of Graz and Vienna [Ranzinger & Gleixner 1997], and in Australia to develop a 3D GIS model for the city of Adelaide [Kirkby et al., 1997]. Notable work on city modeling has also been performed by the MIT group of Seth Teller, focusing mostly on image capturing, sensor calibration, and scene modeling using specially developed equipment like the Argus camera and the roaming platform of Rover [Antone & Teller, 2000]. Argus is a high-resolution digital camera mounted on a small mobile platform and wheeled around campus. It incorporates specialized instrumentation to estimate the geolocation of exposure station and camera orientation parameters for each image acquired. Rover is a controlled vehicle used to acquire geo-referenced video images of interiors and exteriors.

These VR models of urban scenes are photorealistic: they provide views of the world very similar to the ones we would perceive if we were to roam the scene, sometimes even to the point of including graffiti on the walls. However, these models are not tempo-realistic: the real world is in flux, yet these models represent only a single instance of the scene, namely the moment when the images used to create them were actually collected. Considering the high cost to actually build such models, their updating is rarely a priority, unless of course specific information (e.g. the demolition of an important building) makes it necessary to update a small part of the database. Furthermore, it is often remarked that VR models feel empty, failing to incorporate the movement of vehicles and people. This lack of temporal validity has hindered the use of virtual models as convenient interface to spatial databases, even though they convey geospatial information and their expressive power is of tremendous value to the communities that use geospatial information in everyday activities.

Geosensor networks force us to re-evaluate whether this rather static visualization approach is actually adequate. The challenge we face is to incorporate the temporal aspect into VR models, thus supporting theior evolution to Virtual GeoReality (VGR) models. Our vision of a VGR model is characterized by two important properties that are not offered by current VR models:

- *automated updates* to capture the current state of the scene they depict, and the
- ability to communicate the *temporal evolution* of their content (e.g. changes in the façade of a building, the movement of vehicles, and the spread of a fire within a scene).

Thus a VGR model is much more than a display of up-to-date geospatial information. It should be perceived as a novel form of a portal to spatiotemporal information as it is captured by distributed sensors: the evolution of an object may be captured by numerous distributed sensors. At the same time the VGR model also provides the integrating medium to link all these sensors into a network: by identifying common objects in their feed (e.g. the same car in different instances, or the same building) we can link the feeds of multiple sensors that otherwise would only share a common temporal reference. In this manner Virtual GeoReality models reflect a merging of virtual and spatiotemporal models. Through this merger each field can be infused by the advantages of the other, introducing the dimension of time and enabling complex spatiotemporal analysis in the use of VR models, and enhancing GIS with the superb communication capabilities of VR models.

6. OUTLOOK AND OPEN ISSUES

Geosensor networks are a rapidly evolving multidisciplinary field that challenges the research areas involved to integrate new techniques, models and methods that are often not found in their classical research agendas. Interdisciplinary workshops like the first Geo Sensor Networks meeting are an important step towards providing an exchange forum for this newly emerging community. Due to the large overlap of research challenges but varying backgrounds in the different domains, such workshops can be a fruitful opportunity for collaborations. During several panel discussions, open issues were discussed.

One of the prominent open issues using sensor networks today is the issue of sensor data privacy. With the requirements to design ultra-light wireless communication protocols for small-form devices, not much room is left for advanced encryption schemes. A related issue is the need for authentication of sensed data. If sensor networks are deployed in security sensitive areas, built-in mechanisms need to be available to provide for such data authentication. A third open issue is data quality. Mechanisms need to assure that defective or incorrectly calibrated sensors are excluded from the computation, and that calibration is established individually as well as collectively before deployment and also continuously later on. Today, many research efforts in sensor networks are conducted under assumptions derived from the constraints of current hardware platforms such as the Berkeley motes. Many of these assumptions such as using radio broadcasting as communication modality or restricted battery life might not be valid in a few years, and these assumptions might change completely.

ACKNOWLEDGEMENTS

We would like to acknowledge the input of D. Goldin, I. Cruz, M. Egenhofer, A. Howard, A. Labrinidis, S. Madden, S. Voisard and M. Worboys in summarizing the workshop findings. The work of the authors is supported by the National Imagery and Mapping Agency through NURI Award NMA 401-02-1-2008; the workshop of Dr. Stefanidis is further supported by the National Science Foundation through grant ITR-0121269. Finally, we would like to acknowledge the National Science Foundation for supporting the GSN workshop through grant EIA-9876707.

REFERENCES

Ailamaki A., C. Faloutsos, P. Fischbeck, M. Small, and J. VanBriesen, 2003. An Environmental Sensor Network to Determine Drinking Water Quality and Security, *SIGMOD Record,* 32(4), pp. 47-52.

Antone M. and S. Teller, 2000. Automatic Recovery of Relative Camera Rotations for Urban Scenes, *Proceedings of CVPR*, Vol. II, pp. 282-289.

Brenner C., 2000. Towards Fully Automatic Generation of City Models, *Int. Arch. of Photogrammetry & Remote Sensing*, Vol. 33(B3/1), pp. 85-92.

Chen A., R. Muntz, S. Yuen, I. Locher, S. Park, and M. Srivastava, 2002. A Support Infrastructure for the Smart Kindergarten, *IEEE Pervasive Computing,* 1(2), pp. 49-57.

Day A., V. Bourdakis, and J. Robson, 1996. Living with a Virtual City, *Architectural Research Quarterly*, 2, pp. 84-91.

Gruen A. and X. Wang, 1999. CyberCity Modeler, a Tool for Interactive 3D City Model Generation, *Photogrammetric Week'99*, D. Fritsch and R. Spiller (Eds.), Wichmann Verlag, Heidelberg, pp. 317-327.

Haala N. and C. Brenner, 1999. Virtual City Models from Laser Altimeter and 2D Map Data, *Photogrammetric Engineering & Remote Sensing*, 65(7), 787–795.

Hellerstein J., W. Hong, and S. Madden, 2003. The Sensor Spectrum: Technology, Trends, and Requirements, *SIGMOD Record,* 32(4), pp. 22-27.

Howard A., M. Mataric, and G. Sukhatme, 2003. From Mobile Robot Teams to Sensor/Actuator Networks: The Promise and Perils of Mobility, downloadable from www.spatial.maine.edu/~gsn03/program.html

Juang P., H. Oki, Y. Wang, M. Martonosi, L. Peh, and D. Rubenstein, 2002. Energy-Efficient Computing for Wildlife Tracking: Design Tradeoffs and Early Experiences with ZebraNet, in *Proc. Intl. Conf. On Architectural Support for Programming Languages and Operating Systems (ASPLOS-X),* San Jose, CA, pp. 96-107.

Kirby S., R. Flint, H. Murakami, and E. Bamford, 1997. The Changing Role of GIS in Urban Planning: The Adelaide Model Case Study, *International Journal for Geomatics*, 11(8), pp. 6-8.

Lin C., C. Federspiel, and D.M. Auslander, 2002. Multi-Sensor Single Actuator Control of HVAC Systems, in *Proc. Intl. Conf. For Enhanced Building Operations,* Austin, TX.

Mainwaring A., J. Polastre, R. Szewczyk, and D. Culler, 2002. Wireless Sensor Networks for Habitat Monitoring, Technical Report IRB-TR-02-006, Intel Laboratory, UC Berkeley.

Pister K. et al., 2002. 29 Palms Fixed/Mobile Experiment, robotics.eecs.Berkeley.edu/~pister/29Palms0103/.

Ranziger M. and G. Gleixner, 1997. GIS-Datasets for 3D Urban Planning, *Computers, Environments & Urban Systems*, 21(2), pp. 159-173.

Databases and Sensor Networks

Querying Asynchronously Updated Sensor Data Sets under Quantified Constraints

Lutz Schlesinger

University of Erlangen-Nuremberg
Department of Database Systems
Martensstr. 3, 91058 Erlangen, Germany
schlesinger@informatik.uni-erlangen.de

Wolfgang Lehner

Dresden University of Technology
Database Technology Group
01062 Dresden, Germany
lehner@inf.tu-dresden.de

ABSTRACT

Sensor databases are being widely deployed for measurement, detection and surveillance applications. The resulting data streams are usually collected in a centralized sensor database system holding the data belonging to the same subject or the same location. It is common practice to evaluate the data only locally. To discover new results, modern requirements imply to combine data at different sensors focussing on different subjects. The natural way of realizing a homogenous view and a centralized access point to a distributed set of database systems is the concept of a federated or a data warehouse system. However, the tremendously high data refresh rate disallows the usage of these traditional replicated approaches with its strong consistency property. Instead we propose a system that handles packaged updates to reduce the replication frequency and deals with weak consistency properties. Such a system may be based on grid computing technology: This technique mainly used in the "number crunching" area provides an alliance of autonomous local computing systems working together to build a virtual supercomputer. In this paper we combine the grid idea with sensor database systems. A query to the system is sent to one of the grid members, where each node holds cached data of other grid members. Due to the nature of sensor database applications, cached information may become stale fast. In order to speed up query processing the access to outdated data may sometimes be acceptable if the user has knowledge about the degree of staleness or the degree of inconsistency if unsynchronized data are combined for a global view. The contribution of this paper is the presentation and discussion of a model describing inconsistencies in grid organized sensor database systems.

1. ARCHITECTURE OF A GRID SENSOR DATABASE ENVIRONMENT

For years research and industry focus on the efficient analysis of physical data (e.g. weather, traffic, water, soil) or statistical data (e.g. income, structure of population). The data sets are typically collected in sensor database systems ([BoGS01]) and can be locally evaluated. Modern requirements intend to combine data of different sensor databases focussing different subjects like environmental or economical data to discover novell pattern in the data. For global evaluations two approaches are well

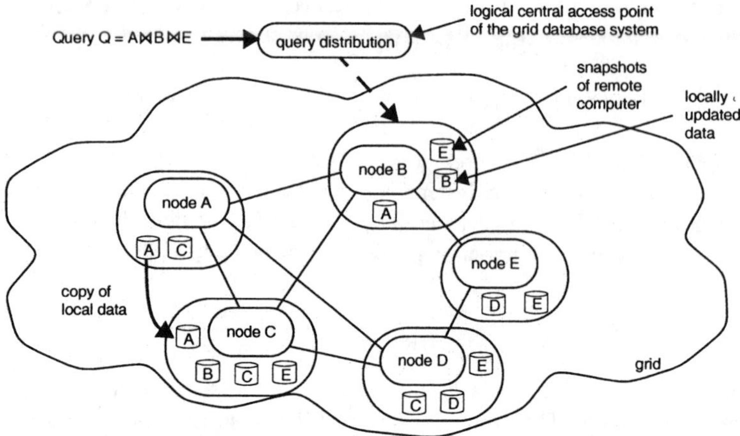

Figure 1: Logical architecture of the discussed grid sensor database scenario.

known: The federated approach ([ÖzVa91]) and the data warehouse approach ([Inmo96], [Kimb96]).

With regard to the usage in a sensor environment they suffer from the strong consistency property which cannot be maintained due to the high data refresh rate. To get these problems under control, we propose a combination of sensor database systems with the concept of *grid organized computing structures* ([FoKe99]). The main idea of a grid is the alliance of autonomous local computing systems to build a single virtual supercomputer. All tasks are centrally managed and distributed to single nodes. The grid technology is well known in the 'number crunching' area (the Globus Project; http://www.globus.org/) but it is not widely used for 'data crunching' where local nodes are represented by database systems. Research in the combination of grid and database technology are manifold and for example coordinated by the *Database Access and Integration-Services* forum (DAIS; http://www.cs.man.ac.uk/grid-db/).

Regarding the described problem we propose the following architecture (figure 1). The grid consists of autonomous nodes, which are connected by a network. Each node collects, stores and updates sensor data in a local sensor database. Since sensor data permanently change their values or add new values to already monitored data, we introduce the concept of recording periods. After finishing a recording period the existing data are frozen and a new recording period starts. A recording period comprises a data set and is finished either after a certain time intervall, a maximum amount of recorded data or on demand from another member of the grid requesting the most up-to-date information. In comparison to the federated approach, our concept avoids permanent data synchronization. In the opposite, the packaged updates do not guarantee the strong consistency property, what might be easier understood from another perspective: An update of a local snapshot during query processing is time and cost consuming regarding the communication and data transfer costs. If a fallback to a local existing but stale snapshot is possible from a user (or application) point of view, the user gives up the highest consistency in two dimensions: The snapshots are not the most current ones and the locally cached snapshots need not to have the same valid time. The time difference between the current time and the time point of the selected snapshots should be quantified by a metric. The user may integrate this metric in the specification of a query to control the maximum inconsistency. Moreover the metric together with other parameters (e.g. communication cost) can be used by the system to distribute the query to one of the grid nodes.

Contribution and Structure of the Paper

In this paper we quantify the metric and propose a general model for dealing with inconsistencies in our grid organized sensor database system. Section 2 formalizes the grid sensor database system. Our *SCINTRA (Semi-Consistent INtegrated Time-oriented Replication Approach*) data model is introduced in section 3 and 4. Section 3 presents the data structure concept, while section 4 discusses several operations. In section 5 an example illustrates the introduced model. An overview of related work is given in section 6. The paper closes with a conclusion.

2. SYSTEM MODEL

In this section we introduce the system model of the grid sensor database system as shown in Figure 2.

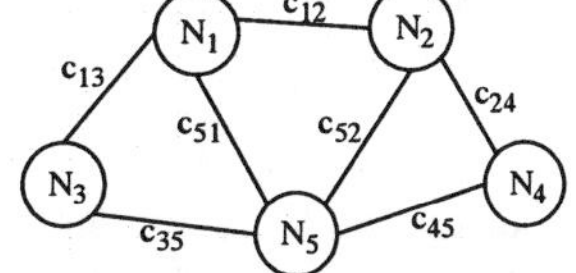

Figure 2: Example of a grid.

Definition 1: Grid database system

A grid database system is a tuple (N, V, C) with

- $N = \{ N_1, N_2, ..., N_n \}$ as a list of nodes N_i
- $V = \{ V_{ij}^* \}$ and $V_{ij} = (N_i, N_j)$ as a list of connections between nodes N_i and N_j $(N_i, N_j \in N)$
- $C = \{ C_{ij}^* \}$ as communication costs between nodes N_i and N_j $(N_i, N_j \in N$ and $(N_i, N_j) \in V)$.

It is assumed that the structure of the network and the underlying network protocol allows the communication between every two nodes of the grid. From a logical point of view the network is completely mashed. On the physical layer not every node may be directly connected with each other, which is represented by the communication costs.

Definition 2: Communication cost between two nodes

The communication cost C_{ab} between two nodes N_a and N_b $(N_a, N_b \in N)$ is defined as $C_{ab} = \min(\Sigma\, c_{ij})$ with $\exists(l_1, l_2, ..., l_m)$ with $l_1 = a$, $l_m = b$, $i = k$ and $j = k+1$ with $1 \leq k \leq m-1$ and $V_{ij} \in V$.

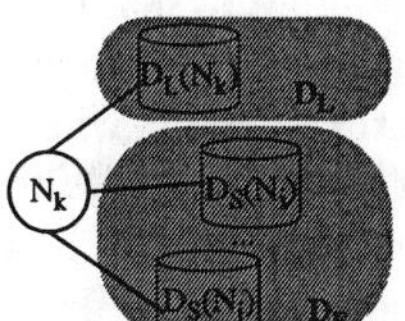

Figure 3: Data sets residing at a grid node.

According to the motivation each node holds locally maintained data D_L and a collection of foreign snapshots D_F.

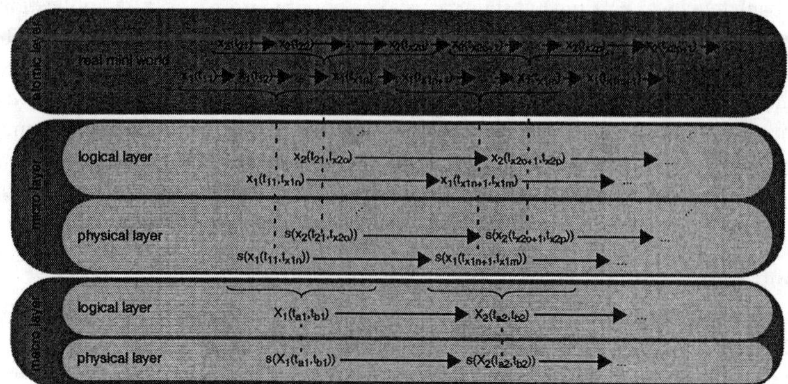

Figure 4: Data structure at a single node of the grid.

Figure 3 illustrates the situation. The structure and the content of the data is discussed in detail in section 3.

Definition 3: Data sets residing at a grid node

The data set of a node N_k ($N_k \in N$) is a tuple (D_L, D_F) with local data $D_L(N_k)$ and foreign data $D_F(N_k) = \{ (D_S(N_i))^* \}$ with $D_S(N_i) = D_L(N_i)$ and $D_L(N_K) \subseteq \bigcup_{i=1, i \neq k} D_S(N_i)$ $(i \neq k)$.

3. DATA STRUCTURE CONCEPT

In our model, we distinguish between three layers at a single node of the grid as illustrated in figure 4. The atomic layer being discussed in detail in section 3.1 reflects data sets of the real world. For example, these data are sent from a sensor. At the next layer, the micro layer (section 3.2), the atomic elements are grouped to micro objects. Finally, several micro objects of a single node form macro objects (section 3.3). In section 3.4 the focus is turned away from a local view to a global view: As mentioned in the introduction, snapshots of local data are sent to other members of the grid. The structure at the node level is discussed in this section.

3.1 Atomic layer

Starting with a look at the real world, sensors measure (physical) signals like temperature, bank account or sales and send the data to a sensor database, where the data are collected and made available for applications to analyze the data. Each such a (sensor) object monitors atomic data consisting of a timestamp and a value, e.g. the temperature at 2003-01-23 13:00 is 20°F.

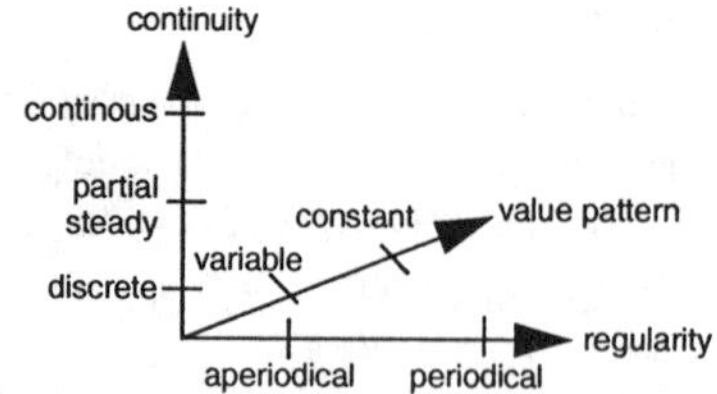

Figure 5: Classification of real objects.

Definition 4: Atomic object

An atomic object is a tuple (x,t,d) consisting of a time stamp t and a data value d, which is monitored by a single sensor with unique identifier x. The short form x(t) = d is defined to denote d.

Although a single physical sensor is possibly capable of measuring different objects, the term "sensor" is used in the context of this paper on a logical level to identify real objects. Each real object may be classified into several dimensions, which are shown in figure 5 and discussed in the rest of this section. We use an extension of the classification made by [TCG+93].

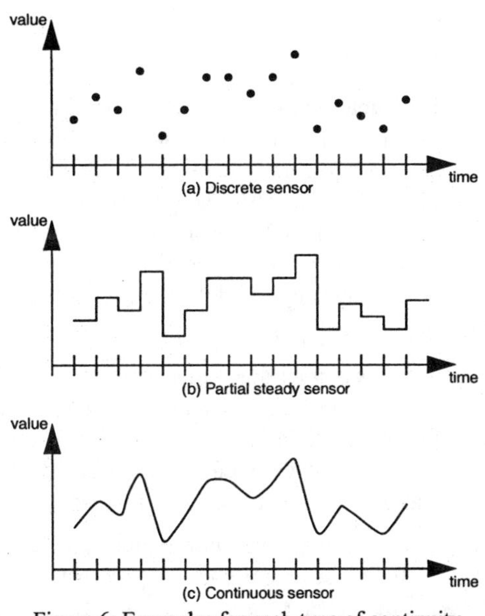

Figure 6: Examples for each type of continuity.

Sensor classification according to the continuity

A first classification is based on the continuity property of the measured real object. The data may be discrete, partial steady or continous although the data sets sent by the sensor are always discrete tuples in the form of (x,t,d).

The first category, discrete data $x(t_i)$, is only 'steady' at the time point t_i. For each timestamp t_i a valid value d_i is defined. Otherwise ($]t_{i-1}; t_i[$ and $]t_i; t_{i+1}[$) x(t) is undefined as shown in figure 6a. Examples of discrete data sets are sales, turnover, number of calls, produced articles or accidents, daily working time, etc.

Partial steady data sets (figure 6b) are characterized by a finite number of salti, i.e. data are steady in the interval $[t_i; t_{i+1}[$. The number of employees or inhabitants, stock, price, tax or bank account are some examples.

The last category are continous data with a continous pattern of the data values and without any salti (figure 6c). It can be described by a mathematical function using the measured discrete data as sampling points, otherwise the sensor is of the type partial steady. Examples are temperature, blood preasure or water consumption.

Sensor classification according to the regularity

Another classification is based on the regularity property, i.e. the distance between two time stamps of a single object x. If $t_{i+1} - t_i = const$ for all i, then the regularity is periodical. Examples for this case are daily sales or the daily value of the stock exchange. Otherwise the bank account is an example for aperiodical regularity.

Sensor classification according to the value pattern

The classification according to the value pattern is the final classification direction. The values between two timestamps t_i and t_{i+1} are either constant or variable. An example for the first one is the bank account, for the second one is the temperature.

3.2 Micro Layer

Atomic objects are events in the real world. At the micro layer, which is discussed in this section, a set of atomic objects is grouped together. A *micro object* is formed after monitoring a number of atomic objects, a time interval or on demand. The group consists of at least a single atomic object and gets timestamps representing the time minimum and time maximum. A function f may be defined to describe the value patterns of the real object, because the monitored atomic objects are only discrete. All this is done on a logical layer and formalized in the following definition.

Definition 5: Logical Micro Object (LMiO)

Represent x a monitored object and $x(t_1), ..., x(t_n)$ a sequence of atomic objects of x with $t_a = \min\{t_1, ..., t_n\}$ and $t_b = \max\{t_1, ..., t_n\}$. Then the logical micro object $LMiO(x,[t_a;t_b])$ is a tupel $(x(t_a,t_b), c, f_{x(t_a, t_b)}(z))$ with $x(t_a,t_b) = \{x(t_1), ..., x(t_n)\}$, $c \in$ {discrete, partial steady, continous} describing the continuity, a function f describing the value pattern and $t_a \leq z \leq t_b$.

At a single node the monitored atomic objects are exactly assigned to a single LMiO. This causes the non-overlapping of two different LMiOs (*LMiO Container Property,* definition 6).

Definition 6: Container Property of two Logical Micro Objects (CP-LMiO)

Represent $LMiO_1(x,[t_{a1};t_{b1}])$ and $LMiO_2(x,[t_{a2};t_{b2}])$ two logical micro objects of the same sensor x. Then the following container properties hold:

- $LMiO_1 \cap LMiO_2 = \varnothing \Leftrightarrow [t_{a1};t_{b1}] \cap [t_{a2};t_{b2}] = \varnothing$
- $LMiO_1 = LMiO_2 \Leftrightarrow [t_{a1};t_{b1}] = [t_{a2};t_{b2}]$

The physical representation of the LMiO stored in the sensor database is the *physical micro object (PMiO)*.

Definition 7: Physical Micro Object (PMiO)

The physical micro object $PMiO(x,[t_a;t_b])$ of a logical micro object $LMiO(x,[t_a;t_b])$ is defined as $s(LMiO(x,[t_a;t_b]))$.

3.3 Macro Layer

On the next layer, the macro layer, several micro objects are grouped to macro objects. In content and schema, the grouped micro objects need not be the same. The schema of the macro object is built by schema integration of the schemas of the underlying micro objects ([ShLa90]). Missing instance values of the micro objects are set to NULL. The *logical macro object (LMaO)* forms a container for a set of logical micro objects. The physical representation is the *physical macro object (PMaO)*, which is sent to other nodes of the grid.

Definition 8: Logical Macro Object (LMaO)

A logical macro object is defined as

$$LMaO(X, [t_A;t_B]) = \bigcup_{i=1}^{M} LMiO(x_i, [t_{ai};t_{bi}])$$

with $X = \{x_1,...,x_M\}$ as a set of monitored objects, $LMiO(x_i,[t_{ai};t_{bi}])$ as logical micro objects, $x_i \neq x_j$ for all $i,j \in \{1, ...,M\}$ $(i \neq j)$ and $t_A = \min\{t_{a1},t_{a2},...,t_{aM}\}$ and $t_B = \max\{t_{b1},t_{b2},...,t_{bM}\}$.

Definition 9: Container property of two logical macro objects (CP-LMaO)

Represent $LMaO_1(X,[t_{A1};t_{B1}])$ and $LMaO_2(X,[t_{A2};t_{B2}])$ two logical macro objects. Then the following container properties hold:

- $LMaO_1 \cap LMaO_2 = \emptyset \Leftrightarrow [t_{A1};t_{B1}] \cap [t_{A2};t_{B2}] = \emptyset$
- $LMaO_1 = LMaO_2 \Leftrightarrow [t_{A1};t_{B1}] = [t_{A2};t_{B2}]$ and
  $\forall\, LMiO_i \in LMaO_1\ \exists\, LMiO_j \in LMaO_2: LMiO_i = LMiO_j$
  $\forall\, LMiO_i \in LMaO_2\ \exists\, LMiO_j \in LMaO_1: LMiO_i = LMiO_j$

Definition 10: Physical Macro Object (PMaO)

The physical macro object $PMaO(X,[t_A;t_B])$ of a $LMaO(X,[t_A;t_B])$ is defined as $S(LMaO(X,[t_A;t_B]))$.

The difference between logical objects (LMiO, LMaO) and physical objects (PMiO, PMaO) is the following: In the grid a logical object exists exactly once, while the physical representation may exist several times. As illustrated in figure 7, for a single LMaO there are several physical representations, whereby a physical representation of a logical object does not reside at every node. As the atomic objects are typically collected at exactly one node N and a sequence of logical macro objects hold data of the same sensors, N is called the *data source*.

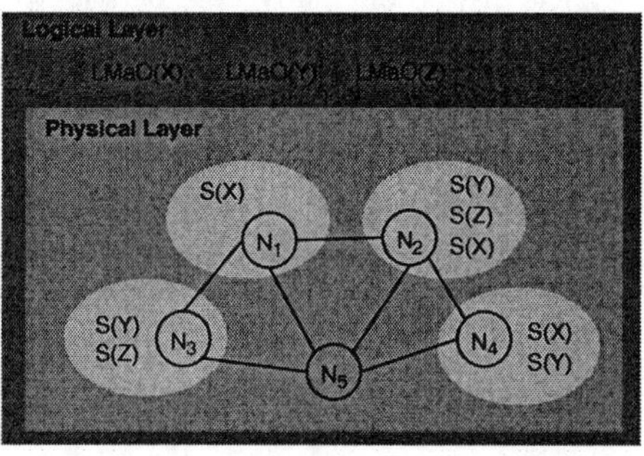

Figure 7: Example for LMaOs and PMaOs.

3.4 Grid Layer

At the *grid layer* a node of a grid collects the atomic objects and stores them in a database denoted as the *source or collecting database* D_s (figure 8). The monitored atomic objects

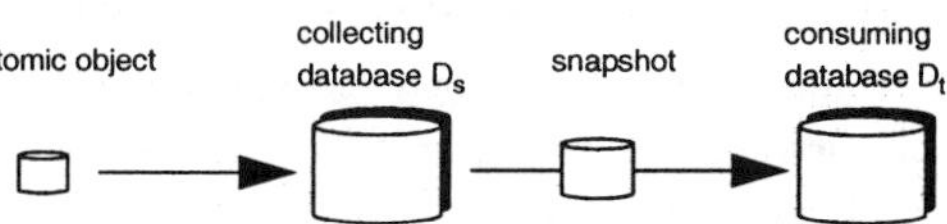

Figure 8: Illustration of objects in the grid.

are grouped to physical macro objects which form the local data D_L. A copy of a PMaO is sent as a snapshot to other nodes of the grid, the *target or consuming database* D_t. As explained in the introduction, the new PMaO is stored in addition to already existing PMaOs from the same database to achieve a better query quality by user specifications. All PMaOs from foreign nodes form the foreign data D_F.

The time of storing the data locally is the transaction time. The valid time is either the upper time bound of the PMaO t_B or the transaction time if no other time is known. For each node the stored snapshots and their valid time can be illustrated in an object-time-diagram as shown in figure 9.

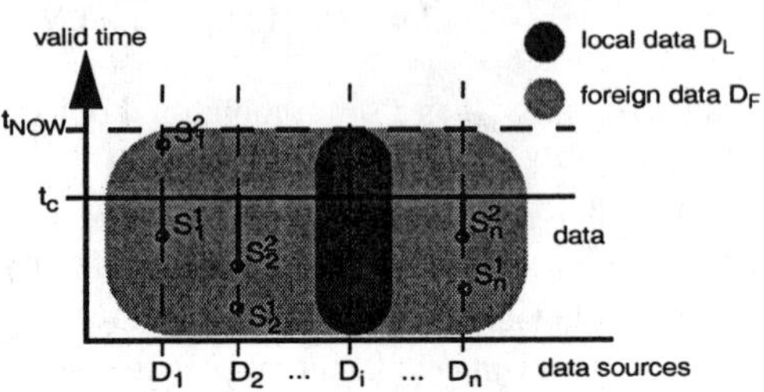

Figure 9: Example for a object-time-diagram at a single node.

4. DATA MANIPULATION CONCEPT

The layer dependent operations defined on the introduced data structure concept are discussed in this section. Section 4.1 starts with the grid layer, where operations at a single node are presented. The following sections deal with the other layers. The complete discussion almost considers logical objects because physical objects are only representations of logical objects.

4.1 Operations at the Grid Layer

At the grid layer general operations executed on a single node concern operations on the collecting and consuming database. The most important operations are insertion, deletion and distribution of objects:

- *Insertion of a new data source*
  This operations adds a new data source of a foreign node so that the arriving physical macro objects are added to the set of foreign data D_F. The schema of the new data source has to be integrated with the already existing local schema and is done by well known strategies ([ShLa90]).
- *Deletion of a data source*
  The reverse operation is the deletion of a data source implying the removal of all physical macro objects existing in D_F. The integrated schema is reduced by the schema of the removed data source.
- *Distribution of a physical macro object*
  As already mentioned, a physical macro object PMaO is sent from the collecting database to other consuming databases. For that, a distribution operation may be defined. A request operation is introduced for requesting the newest PMaO of a collecting database.

4.2 Operations at the Macro Layer

Operations on macro objects are defined at the macro layer, where a set of macro objects of different data sources is given as illustrated in figure 9. If from a user point of view a selection of outdated snapshots is acceptable, algorithms for a parameterized selection of snapshots and operations between the selected snapshots have to be defined and are introduced in the rest of this section.

Historic Cut

In a classical database middleware approach the latest snapshots are selected and joined. These are the closest snapshots to the horizontal line t_{NOW} reflecting the current time in the object-time-diagram. In our approach, where an arbitrary number of macro objects of the same data source exists, a selection at any time t_c ($t_c \leq t_{NOW}$), called the *historic cut*, is possible. In the ideal case all snapshots are valid at the same time (figure 10a) while in our approach the snapshots may have different valid times why the connecting line t_c is not any more a straight line (figure 10b). The curve reflects the inconsistency relating to the different valid times. A metric to quantify the inconsistency is defined in the following. Algorithms for selecting the snapshots under consideration of the metric and the age of the snapshots are discussed in [ScLe02].

Quantifying the Inconsistency

Basically, the inconsistency metric considers the time and the data change rate of a set of snapshots. Disregarding the second aspect for a moment, the time inconsistency for a single data source is defined as the distance between the valid time of the snapshot and the time point of the historic cut. The inconsistency I is defined on the basis of the L_p-metric.

Definition 11: Simple Inconsistency Formula

Represent t_c a historic cut and S_j^i a snapshot. For k selected snapshots the inconsistency I is defined as

$$I = \sqrt[p]{\sum_{i=1}^{k} (\mathrm{time}(S_i^j) - t_c)^p}$$

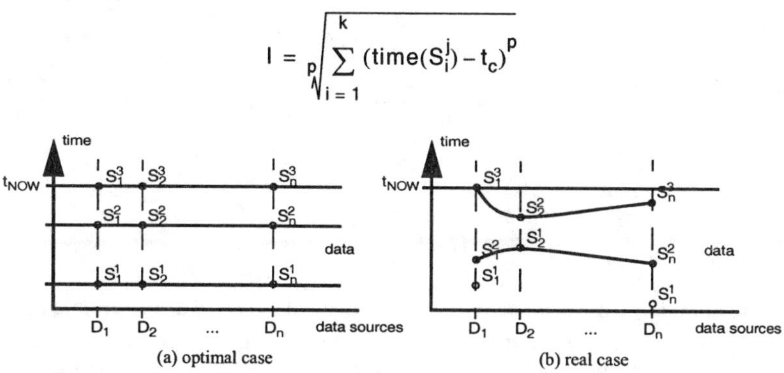

Figure 10: Selection of snapshots in the optimal case and in comparison to the real case.

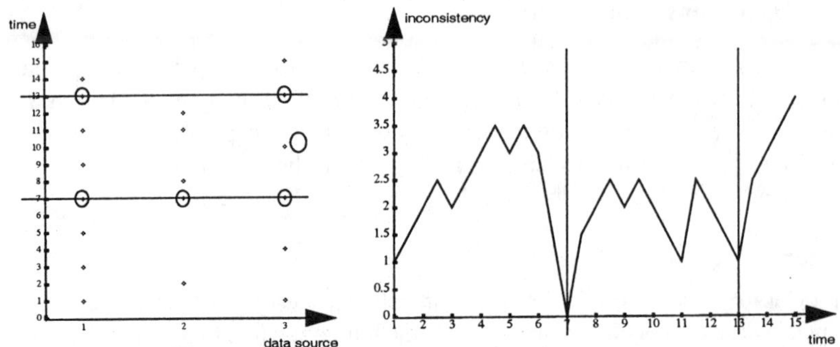

Figure 11: Example for the selection of historic cuts and the resulting inconsistency curve.

The example of figure 11 shows snapshots of three data sources in an object-time-diagram (figure 11 left) with $p = 1$. According to the formula the inconsistency at point 13 is 1 and 0 at point 7 (figure 11 right). This simple example illustrates the conflict between age and inconsistency: At point 7 the inconsistency is lower than the inconsistency at time point 13, but the age is much higher.

While the above inconsistency formula only consists of the distance in time, the formula is now extended to take a data change rate into consideration. For each data source i we introduce a data change rate Δd_i, which reflects the data changes of the macro and the underlying micro objects. This signature considers existential changes (insertions and deletions of objects) and has values between 0 and 1 (0-100%).

Definition 12: Reusage Rate

Represent Δd_i the data change rate of data source i. Then the reusage rate ρ at a time point t with t_S as the time point of the global oldest snapshot is defined as

$$0 < \Delta d_i \leq 0,5 : \rho(t) = \frac{-1}{|t-(t_{NOW}-t_S)-1|^{\frac{1}{2\Delta d_i}-1}} \bullet \left(\frac{1}{t_{NOW}-t_S} \bullet (t-(t_{NOW}-t_S))+1\right)+1$$

$$0,5 \leq \Delta d_i < 1 : \rho(t) = \frac{1}{(t+1)^{\frac{-1}{2(\Delta d_i-1)}-1}} \bullet \left(\frac{-1}{t_{NOW}-t_S} \bullet t+1\right)$$

At time point t_{NOW} ρ has a value of 100% independently of Δd_i and at time point t_S the value is 0. The value pattern between these points is determined by Δd_i and illustrated in figure 12. The combination of the distance in time and the reusage degree leads to an extended inconsistency formula.

Definition 13: Extended Inconsistency Formula

Denote α_k weights of each single data source and $S_k^{j_k}$ the selected snapshots. Then the extended inconsistency formula is defined as

$$I(\text{time}(S_1^{j_1}), \ldots, \text{time}(S_k^{j_k})) = \sum_{k=1}^{n} (\alpha_k \bullet |(\text{time}(S_k^{j_k}) - t_c)| \bullet (1 - \rho_k(\text{time}(S_k^{j_k}))))$$

Operations

After selecting at most one single snapshot for each data source, operations on the macro objects may be defined. As a macro object is only a container for a set of micro objects almost all operations are defined on a single micro object or between two or more micro objects (section 4.3). An exception are the selection of micro objects, which results in a subset of the micro objects, which satisfy the restriction condition and belong to a single macro object.

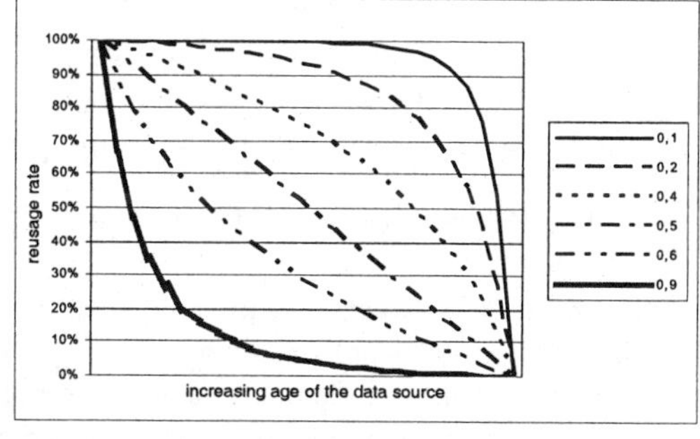

Figure 12: Dependency of reusage degree and age.

4.3 Operations at the Micro Layer

At the micro layer operations on a single micro object or between two micro objects are defined. Mostly the operations are realized by operations on atomic objects, which are discussed in section 4.4. The considered operations are the standard operations (selection, projection, join, unary and binary operations, operations on sets and aggregation operations) with its special focus on sensor data sets and under the consideration of the sensor classification (section 3.1).

A selection operation returns a set of atomic objects of a single micro object, if the selection condition holds, otherwise the empty set. The projection operation and the unary operation return a micro object containing the number of atomic objects but having different values.

Definition 14: Selection operation on a micro object

The selection operation on a logical micro object $LMiO = (x,[t_a;t_b])$ with the selection condition scond is defined as $\sigma_{scond}(LMiO) = (x',[t_{a'};t_{b'}])$ with $x' = (x'(t_{a'},t_{b'}), c, f_{x'(t_{a'},t_{b'})}(z'))$ and with

- $x'(t_{a'},t_{b'}) = \{\sigma_{scond}(x(t_i))^*\}$ and $t_{a'} = \min \{\sigma_{scond}(x(t_i))^*\}$ and $t_{b'} = \max \{\sigma_{scond}(x(t_i))^*\}$ for all $x(t_i) \in x$
- $$f_{x'(t_{a'},t_{b'})}(z') = \begin{cases} f_{x(t_a,t_b)}(z) & \forall x(t_i) \in x \cap x' \\ & i \in [t_{a'};t_{b'}] \text{ and } t_{a'} \leq z' \leq t_{b'} \\ \text{undef} & \text{else} \end{cases}$$

Definition 15: Projection operation on a micro object

The projection operation on a logical micro object LMiO = $(x,[t_a;t_b])$ with the projection condition pcond is defined as $\pi_{pcond}(LMiO) = (x',[t_a;t_b])$ with $x' = (x'(t_a,t_b), c, f_{x(t_a,t_b)}(z))$ and $x'(t_a,t_b) = \{\pi_{pcond}(x(t_i))^*\}$ for all $x(t_i) \in x$ and $t_a \leq z \leq t_b$.

Definition 16: Unary operation on a micro object

The unary operation on a logical micro object LMiO = $(x,[t_a;t_b])$ is defined as

- $\theta(LMiO) = (x',[t_a;t_b])$ with $\theta \in \{-, abs, sign\}$, $x' = (x'(t_a,t_b), c, f_{x'(t_a,t_b)}(z))$ and $x'(t_a,t_b) = \{\theta(x(t_i))^*\}$ for all $x(t_i) \in x$
- $\theta_v(LMiO) = (x',[t_a;t_b])$ with a numerical variable v, $\theta \in \{<,>,\leq,\geq,=,\neq\}$, $x' = (x'(t_a,t_b), c, f_{x(t_a,t_b)}(z))$ and $x'(t_a,t_b) = \{\theta_v(x(t_i))^*\}$ for all $x(t_i) \in x$

The new function $f_{x'(t_a,t_b)}(z)$ describes the new value pattern in the interval $t_a \leq z \leq t_b$.

A join operation between two micro objects means a joining of the respective atomic objects, which would result in a 6-tuple. As the unit in our model is an atomic object consisting of three elements the join operation can only be defined in combination with a binary operation to reduce the 6-tuple to a 3-tuple. Therefore, the join operation is identical with the binary operation called *SCINTRA-Join*.

The join of two micro objects includes the problem of which atomic objects may be selected, because the time points and the type of the sensors may be different. The problem is illustrated in figure 13. The micro object shown in the upper part reflects a continous sensor while the other sensor is discrete. The bullets mark monitored data (atomic objects). The selection of two atomic objects depends on the join semantic, which may be for example most recent, most next, average or exact following the SEQ-model ([SeLR95]). The resulting continuity type depends on the application.

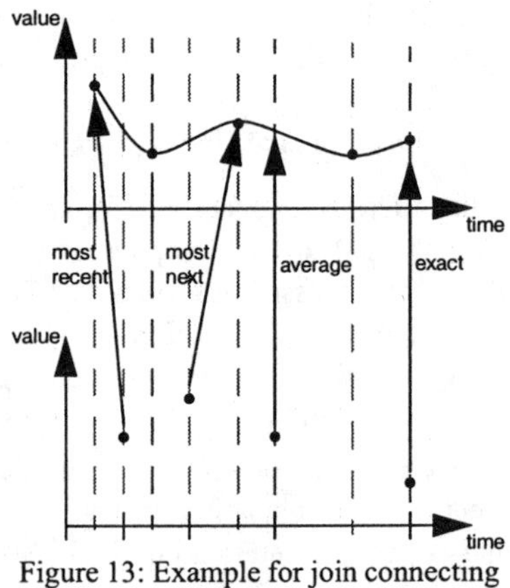

Figure 13: Example for join connecting semantics.

Definition 17: Binary operation of two micro objects (SCINTRA-Join)

The binary or join operation between two micro objects $LMiO_1 = (x_1,[t_{a1};t_{b1}])$ and $LMiO_2 = (x_2,[t_{a2};t_{b2}])$ is defined as

$$LMiO_1 \bowtie_{s,\theta,c} LMiO_2 = LMiO_1\ \theta_{s,c}\ LMiO_2 = (x,[t_a;t_b])$$

with $x = (x(t_a,t_b), c, f_{x(t_a,t_b)}(z))$ and $x(t_a,t_b) = \{(x(t_i)\ \theta_t\ x(t_j))^*\}$ for all $(x(t_i), x(t_j))$ with $x(t_i) \in x_1(t_{a1},t_{b1})$ and $x(t_j) \in x_2(t_{a2},t_{b2})$ satisfying the join semantic s, $\theta \in \{+,-,*,/,min,max\}$ and with resulting time point t ($\min\{t_{a1},t_{a2}\} \leq t \leq \max\{t_{b1},t_{b2}\}$).

Similar to the binary operation the aggregation operation on a micro object produces a single atomic object from a set of atomic objects.

Definition 18: Aggregation operation on a micro object

The aggregation operation on a logical micro object LMiO = $(x,[t_a;t_b])$ is defined as $agg_a(LMiO) = (x',[NULL;NULL])$ with $x' = (agg_a(x(t_a;t_b)), c, undef)$ and $agg \in \{sum, min, max, count, avg\}$ and $a \in \{time, value\}$.

The final operation introduced here is the set operation on two micro objects, which is only defined, if the micro objects contain data from the same sensor.

Definition 19: Set operation on two micro objects

A set operation of two micro objects $LMiO_1 = (x_1,[t_{a1};t_{b1}])$ and $LMiO_2 = (x_2,[t_{a2};t_{b2}])$ is only defined, if $x_1 = x_2 = x$ and $c_1 = c_2 = c$. Then the set operation is defined as $LMiO_1\ \theta\ LMiO_2 = (x,[ta';tb'])$ with $\theta \in \{\cap,\cup,\backslash\}$, $x = (x(t_a,t_b), c, f_{x(t_a,t_b)}(z))$ and $t_a = \min\{t_{a1},t_{a2}\}$ and $t_b = \max\{t_{b1},t_{b2}\}$ and $x(t_a,t_b) = \{(x(t_{a1},t_{b1})\ \theta_t\ x(t_{a2},t_{b2})\}$ and

$$f_{x(t_a,t_b)}(z) = \begin{cases} f_{x_1(t_{a1},t_{b1})}(z) & t_{a1} \le z \le t_{b1} \\ f_{x_2(t_{a2},t_{b2})}(z) & t_{a2} \le z \le t_{b2} \end{cases}$$

4.4 Operations at the Atomic Layer

At the atomic layer the operations introduced in the last section are now realized by using atomic objects of one or two micro objects. A selection operation returns the atomic object if the selection condition holds otherwise the empty set. The projection operation and the unary operation return a modified atomic object.

Definition 20: Selection operation on an atomic object

The selection operation on an atomic object $x(t)$ with the selection condition scond is defined as

$$\sigma_{scond}(x(t)) = \begin{cases} x(t) & \text{if } scond(t) = true \text{ or } scond(d) = true \\ \varnothing & \text{else} \end{cases}$$

Definition 21: Projection operation on an atomic object

The projection operation on an atomic object (x,t,d) with the projection condition pcond is defined as

$$\pi_{pcond}(x, t, d) = \begin{cases} (x, t, d) & \text{if } pcond = [t,d] \\ (x, t, NULL) & \text{if } pcond = [t] \\ (x, NULL, d) & \text{if } pcond = [d] \\ \varnothing & \text{else} \end{cases}$$

Definition 22: Unary operation on an atomic object

The unary operation on an atomic object $x(t) = (x,t,d)$ with $\theta \in \{-, abs, sign\}$ or with a numerical variable ν and $\theta \in \{<,>,\le,\ge,=,\ne\}$ is defined as:

$$\theta(x(t)) = (x, t, \theta(d))$$

$$\theta_v(x(t)) = \begin{cases} (x, t, d) & \text{if } [d\theta\upsilon] = true \\ (x, t, NULL) & \text{else} \end{cases}$$

As explained in section 4.3 the result of a binary or join operation and the result of an aggregation operation of two atomic objects is an atomic object.

Definition 23: Binary operation of two atomic objects (SCINTRA-Join)

The binary operation between two atomic objects $x_1(t_1) = (x_1,t_1,d_1)$ and $x_2(t_2) = (x_2,t_2,d_2)$ is defined as $x_1(t_1)\ \theta_t\ x_2(t_2) = (NULL,t,d_1\ \theta\ d_2)$ with $\theta \in \{+,-,*,/, min,max\}$ as binary operators and the resulting time t.

Definition 24: Aggregation operation of two atomic objects

The aggregation operation on $x([t_1;t_2]) = agg(\{x(t_1),x(t_2)\})$ is defined as

$$agg_a(\{x(t_1), x(t_2)\}) = \begin{cases} \{(x, agg(t_i), d_i)^*\} & \text{for a = time} \\ \{(x, t, agg(d_i))^*\} & \text{for a = value} \end{cases}$$

While the operations defined above result in a single atomic object the set operations introduced below result in a set, which forms a logical micro object.

Definition 25: Set operation of two atomic objects

The set operation between two atomic objects (x,t_1,d_1) and (x,t_2,d_2) of the same sensor x is defined as

$LMiO(x,[t_a;t_b]) = (x,t_1,d_1)\ \theta\ (x,t_2,d_2) = ((\{(x,t_1,d_1), (x,t_2,d_2)\}, c, f_{x(t_a,\ t_b)}(z)), [t_a;t_b])$ with $t_a = \min\{t_1, t_2\}$ and $t_b = \max\{t_1, t_2\}$, with $\theta \in \{\cap,\cup,\backslash\}$, a continuity type c and a function $f_{x(t_a,\ t_b)}(z)$.

5. EXAMPLE

A simple example of a grid is shown in figure 14. It consists of two nodes each holding a set of snapshots. A user sends a query to the grid to join snapshots of two data sources D_1 and D_2. If the query is routed to N_1 then the query can be evaluated at the historic cut t_{c11} resulting in a computed inconsistency of 5. Another way is to transfer the snapshot S_2^2 from N_2 to N_1 with transfer costs $c_{21} = 50$ resulting in an inconsistency of 1 at t_{c12}. If the query is routed to N_2, then a query evaluation is possible, if S_1^3 is copied from N_1 to N_2 with $c_{12} = 20$ and a resulting inconsistency of 1. Table 1 summarizes the facts. If the user specifies an upper bound of the allowed age of the historic cut, the transfer costs and the inconsistency the system is able to decide on which node the query has to be evaluated. Deciding in such a manner the snapshots are also selected. The necessary framework for quantifying the inconsistency is introduced in section 4.2. The computation of the transfer costs follows [ÖzVa91], whereby the number of tuples corresponds to the number of atomic objects.

After choosing a node and a snapshot for each data source operations between the elements of the snapshots are executed corresponding to a specification in the query. At first the micro objects are selected, which are uniquely determined by a sensor identifier and a time interval. The two are parts of the declaration of the query. In the same way the operations on the atomic objects are specified. For example for a join operation the join semantic, the binary operator and the resulting continuity type may be specified. Thus the operators and its parameters defined in the sections 4.3 and 4.4 allow the user to control the query execution to receive the expected result.

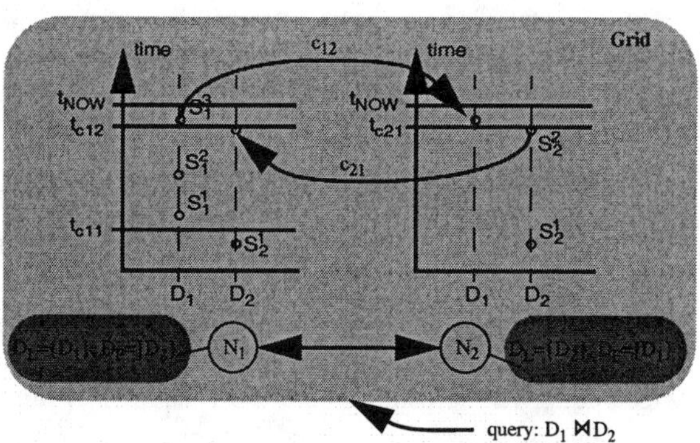

Figure 14: Example for operations on the grid.

Table 1: Overview of computed data in the example

historic cut	transfer costs	inconsistency
t_{11}	0	5
t_{12}	20	1
t_{21}	50	1

6. RELATED WORK

Since our approach is related to many different research areas we classify our materials as follows:

- *mediator and database middleware systems / multidatabase systems*
  Database middleware systems like DataJoiner/Garlic ([IBM01a], [IBM01b]) or mediator systems in general ([Wied92], e.g. TSIMMIS: [IBM01c]) provide a query engine with sophisticated wrapper and query optimization techniques ([RoSc97]). Since those systems including multidatabase systems ([BrGS92], [ShLa90]) they are only considering the current state of a data source, the integration point of different states of data sources from a global point of view is not a subject of interest.
- *temporal database systems*
  Many approaches are dealing with temporal aspects in databases ([TCG+93]). Transaction time and valid time ([JCE+94]) are used to control and manage data from a temporal point of view. Although our approach does not explicitly con-

sider valid time, many ideas from historical data models have influenced our work. Picking the optimal states within a temporal database however, is not discussed so far.

- *replication in distributed database systems*
  Introducing replicas in distributed database systems ([ÖzVa91]) is a general way to speed up local access to data. Since replicas are synchronized immediately in the context of an update operation, our approach is much more related to the concept of database snapshots ([AdLi80]). If a data source supports an incremental update of a snapshot ([LaGa96], [LHM+86]), we may take advantage of this technique. The selection of the "best" set of snapshots from a consistency point of view however, has not been discussed prior to our work.

- *concurrency control in relational database systems*
  Concurrency control reflects an extremely well understood area of database research ([GrRe93]). One way to reduce conflicts between multiple users when concurrently accessing the same data object is to keep multiple and – private – versions of it ([BeGo83], [LuZP01]). However, the corresponding algorithms are considering only a *single* data object and no relationship, i.e. joins, between these objects, so that our problem does not occur.

7. CONCLUSION

The concept of sensor grid database systems allows a member of the grid to cache data from other members of the grid. If from a user point of view a fallback to outdated data is acceptable during query evaluation the query may be faster evaluated but the user gives up highest consistency. In this paper we introduce a model for a sensor grid database system and a consistency framework to quantify the appearing inconsistencies. Our approach is a solid base for the query optimizer to decide on which node a query ought to be evaluated. The model is further helpful for the user to specify how operations on the selected snapshots ought to be executed.

REFERENCES

AdLi80 Adiba M.E. and B.G. Lindsay, 1980. Database Snapshots, in *Proceedings of the 6th International Conference on Very Large Data Bases (VLDB '80)*, Montreal, Canada, October 1-3, pp. 86-91.

BeGo83 Bernstein P. and N. Goodman, 1983. Multiversion Concurrency Control - Theory and Algorithms, in *ACM Transactions in Database Systems 8(1983)4*, pp. 465-483.

BoGS01 Bonnet P., J. Gehrke and P. Seshadri, 2001. Towards Sensor Database Systems, in *Proceedings of the 2nd International Conference on Mobile Data Management (MDM'01)*, Hong Kong, China, January 8-10), pp. 3-14.

BrGS92 Breitbart Y., H. Garcia-Molina and A. Silberschatz, 1992. Overview of Multidatabase Transaction Management, in *VLDB Journal 1(2),* pp. 181-293.

FoKe99 Foster T. and C. Kesselman, C. (ed.), 1999. *The Grid: Blueprint for a New Computing Infrastructure.* Morgan-Kaufmann Verlag, San Francisco, CA.

GrRe93 Gray J. and A. Reuter, 1993. *Transaction Processing - Concepts and Techniques.* Morgan Kaufmann Publishers, San Mateo, CA.

IBM01a 2001. *DataJoiner*, IBM Corp. (electronic version: http://www-4.ibm.com/software/data/datajoiner/)

IBM01b 2000. *The Garlic Project*, IBM Corp., 2001 (electronic version: http://www.almaden.ibm.com/cs/garlic/ homepage.html)

IBM01c 2000. *The TSIMMIS Project.* (electronic version: http://www-db.stanford.edu/tsimmis/ tsimmis.html)

Inmo96 Inmon W.H., 1996. *Building the Data Warehouse.* John Wiley & Sons, NewYork, NY.

JCE+94 Jensen C.S., J. Clifford, R. Elmasri, S.K. Gadia P.J. Hayes and S. Jajodia, 1994. A Consensus Glossary of Temporal Database Concepts, in *SIGMOD Record 23(1)*, pp. 52-64.

Kimb96 Kimball R., 1996. *The Data Warehouse Toolkit.* John Wiley & Sons, NewYork, NY.

LaGa96 Labio W. and H. Garcia-Molina, 1996. Efficient Snapshot Differential Algorithms for Data Warehousing, in *Proceedings of the 22th International Conference on Very Large Data Bases (VLDB'96)*, Bombay, September 3-6, pp. 63-74.

LHM+86 Lindsay B.G., L.M. Haas, C. Mohan; H. Pirahesh and P.F. Wilms, 1986: A Snapshot Differential Refresh Algorithm, in *Proceedings of the 1986 ACM SIGMOD International Conference on Management of Data (SIGMOD'86)*, Washington, D.C., May 28-30, pp. 53-60.

LuZP01 Lu, B.; Q. Zou and W. Perrizo, 2001. A Dual Copy Method for Transaction Separation with Multiversion Control for Read-Only Transactions, in *Proceedings of the 2001 ACM Symposium on Applied Computing (SAC'01)*, Las Vegas, NV, March 11-14, pp. 290-294.

ÖzVa91 Özsu, M. and P. Valduriez, 1991. *Principles of Distributed Database Systems.* Prentice-Hall, Englewood Cliffs, NJ.

RoSc97 Roth, M.T. and P.M. Schwarz, 1997. Don't Scrap It, Wrap It! A Wrapper Architecture for Legacy Data Sources, in *Proceedings of 23rd International Conference on Very Large Data Bases (VLDB'97)*, Athens, Greece, August 25-29, pp. 266-275.

ScLe02 Schlesinger L. and W. Lehner, 2002: Extending Data Warehouses by Semi-Consistent Database Views, in *Proceedings of the 4th International Workshop on Design and Management of Data Warehouses (DMDW'02)*, Toronto, Canada, May 27, pp. 43-51.

SeLR95 Seshadri P., M. Livny and R. Ramakrishnan, 1995. SEQ: A Model for Sequence Databases, in *Proceedings of the 11th International Conference on Data Engineering (ICDE'95)*, Taipei, Taiwan, March 6-10, pp. 232-239.

ShLa90 Sheth A.P. and J.A: Larson, 1990. Federated Database Systems for Managing Distributed, Heterogeneous, and Autonomous Databases, in *ACM Computing Surveys 22(3)*, pp. 183-236.

TCG+93 Tansel A., J. Clifford, S. Gadia, S. Jajodia, A. Segev and R. Snodgrass, 1993. *Temporal Databases*. Benjamin/Cummings Publishing, Redwood City, CA.

Wied92 Wiederhold G., 1992: Mediators in the architecture of future information systems, in: *IEEE Computer 25(2)*, pp. 38-49.

Window Query Processing in Highly Dynamic GeoSensor Networks: Issues and Solutions

Yingqi Xu Wang-Chien Lee

Department of Computer Science and Engineering,
Pennsylvania State University, University Park, PA 16802
E-Mail: {yixu, wlee}@cse.psu.edu

ABSTRACT

Wireless sensor networks have recently received a lot of attention due to a wide range of applications such as object tracking, environmental monitoring, warehouse inventory, and health care [15, 29]. In these applications, physical data is continuously collected by the sensor nodes in order to facilitate application specific processing and analysis. A database-style query interface is natural for development of applications and systems on sensor networks. There are projects pursuing this research direction [13, 14, 25]. However, these existing works have not yet explored the *spatial* property and the *dynamic* characteristics of sensor networks.

In this paper, we investigate how to process a *window query* in *highly dynamic GeoSensor networks* and propose several innovative ideas on enabling techniques. The networks considered are highly dynamic because the sensor nodes can move around (by self-propelling or attaching themselves to moving objects) as well as turn to sleeping mode. There exist many research issues in executing a window query in such sensor networks. The dynamic characteristics make those issues non-trivial. A critical set of networking protocols and access methods need to be developed. In this paper, we present a location-based stateless protocol for routing a window query to its targeted area, a space-dividing algorithm for query propagation and data aggregation in the queried area, and a solution to address the user mobility issue when the query result is returned.

1. INTRODUCTION

The availability of low-power micro-sensors, actuators, embedded processors, and RF radios has enabled distributed wireless sensing, collecting, processing, and dissemination of complex environmental data in many civil and military applications. In these applications, queries are often inserted into a network to extract and derive information from sensor nodes. There are a lot of research efforts aiming at building sensor network based systems to leverage the sensed data to applications. However, most of the existing works are based on design and requirements of some specialized application. Thus, they cannot be easily extended for other applications. To facilitate rapid development of systems and applications on top of sensor networks, building blocks, programming models and service infrastructures are necessary to bridge the gap between underlying sensor networks and upper layer systems and applications.

A database style query interface is natural for development of applications and systems on sensor networks. The declarative, ad hoc query languages used in traditional database systems can be used to formulate queries to exploit various functionality of sensor nodes and retrieve data from the physical world. In deed, database technology, after many years of development, has matured and contributed significantly to the rapid growth of business and industry. Commercial, research, and open-source development tools are available to facilitate rapid implementations of applications and systems on databases. Thus, a query layer on top of the sensor networks will allow database developers to leverage their experience and knowledge and to use existing tools and methodologies for designs and implementations of sensor network based systems and applications.

Sensor databases such as Cougar [25] and TinyDB [13, 14] have been proposed. However, these existing works have not yet exploited the *spatial* property and the *dynamic* characteristics of sensor networks. In this paper, we investigate how to process a spatial *window query* in *highly dynamic sensor networks* (HDSN) and present several innovative ideas on enabling techniques for query processing. The network is highly dynamic because sensor nodes may go to *sleeping* mode to save energy as well as move around by self-propelling or attaching themselves to moving objects (e.g. vehicles, air, water). In addition to the capacities static sensor nodes typically possess (e.g. computation, storage, communication and sensing ability), here we assume that sensor nodes are location-aware via GPS or other positioning techniques [6, 17]. The spatial property of sensor nodes is important since sensor networks are deployed and operated in a geographical area after all. We are particularly interested in window query because it is one of the most fundamental and important queries supported in spatial databases. A window query on sensor database retrieves the physical data falling within specified *query window*, a 2- to 3-dimensional area of interest specified by its user.

There are obviously many new challenges for processing spatial window queries in HDSNs. In this paper, we use the following query execution plan as a vehicle to examine various research issues:

1. Routing the query towards an area specified by the query window;
2. Propagating the query within the query window;
3. Collecting and aggregating the data sensed in the query window;
4. Returning query result back to the query user (who is *mobile*).

Many technical problems need to be answered in order to carry out this plan. For example, how to route the query to the targeted area by taking energy, bandwidth, and latency into account; how to ensure a query reaches all the sensors located within window; how to collect and aggregate data without relying on a static or fixed agent; and how to deal with user mobility. To realize this execution plan, a critical set of networking protocols and access methods need to be developed. Although there is some work investigating either the window query processing or HDSNs, none provides a complete solution for window query processing in a HDSN. We have proposed innovative ideas and enabling solutions. Our proposals prevail for window query processing in HDSNs in the following aspects:

- Sensor nodes are able to make wise query routing decisions without state information of other nodes or the network. The proposed stateless protocol, namely, *spatial query routing* (SQR) enables efficient query routing in HDSNs where the topology frequent changes. Instead of serialized forwarding, pipelining techniques are employed in the protocol to reduce the delay of forwarder selection.
- Queries are propagated inside the query window in an energy-efficient way. The propagation is ensured to cover the whole query window.
- Query results are aggregated in a certain geographical region instead of at some pre-defined sensor node, which adopts well to the dynamics of HDSNs. Query results are processed and aggregated inside the query window, thus the number of transmissions is reduced.
- User mobility is accommodated by utilizing the static property of geographical region as well. The query result is delivered back to the mobile user, even if she moves during query processing.

The rest of this paper is organized as follows. Section 2 presents the backgrounds and the assumptions for our work and discusses various performance requirements. In Section 3, research challenges arising in the context of this study are investigated. Section 4 describes our main designs including spatial query routing algorithm, spatial propagation and aggregation techniques and a strategy for returning the query results back to mobile users. Related work is reviewed and compared with our proposals in Section 5. Finally, Section 6 concludes this paper and depicts future research directions.

2. PRELIMINARIES

In this section, we provide some backgrounds and discuss challenges faced in processing window queries in HDSNs. We first describe the assumptions we use as a basis, followed by a review of HDSNs and window query. At the end, we give a list of performance metrics that need to be considered for evaluating query processing in sensor networks.

2.1 Assumptions

We assume that the sensor network is a pull-based, on-demand network. In other words, the network only provides data of interest upon users' requests. While the types of events and sensed data (e.g. temperature, pressure or humidity) are pre-defined and accessible from the sensor nodes, no sensing or transmission actions are taken by the nodes until the query is inserted into the network. This assumption is based on the fact that most of the sensor networks stay in low power mode in order to conserve energy and prolong the network lifetime. Nevertheless, a push-based network can be emulated by executing a long-running query in an on-demand network. We further assume that users are able to insert their queries from any sensor node, instead of through one or more stationary access points in the sensor networks. Finally, a user, who moves at will, is able to receive the query result back at different locations of the network.

2.2 Highly Dynamic Sensor Networks

Here we characterize the highly dynamic sensor networks (HDSNs). Generally speaking, the sensor nodes in HDSNs have the same functionalities of sensing, computation, communication and storage as the static sensor nodes commonly considered in the literature [1, 7]. Nevertheless, HDSNs also have the following important properties:

- Node Mobility: The sensor nodes in HDSNs are mobile. They may drive themselves by self-propelling (via wheels, micro-rockets, or other means) or by attaching themselves to certain transporters such as water, wind, vehicles and people. With self-propelling sensor nodes, a HDSN is self-adjustable to achieve better area coverage, load balances, lifetime, and other system functionalities. These intelligent sensor nodes can be controlled by the network administrator and adaptable to the queries or commands from the applications. On the other hand, for the sensor nodes attaching to transporters, their moving patterns are dependent on the transporters. The applications may have little control or influence on their movement.
- Energy Conservation: Sensor nodes may switch between *sleeping* mode and *active* mode in order to conserve energy and extend the lifetime of networks. Thus, a sensor node is not always accessible. From the viewpoint of the network, the sensor node joins and leaves the network periodically or asynchronously based on sleeping schedules derived from various factors such as node density, network size, bandwidth contention, etc.

- Unreliable Links and Node Failures: Another factor that contributes to the dynamics of networks is node and communication failures. This has a different impact from energy conservation because the available sensor nodes within the network will continue to decrease.

Sensor nodes with some or all of the above properties form a dynamic sensor network. While nodes sleep, node failure and unreliable communication exist in most sensor networks, here we stress the high mobility of sensor nodes. We argue that the mobility of sensor nodes is essential in a wider range of applications. For example, a sensor network for air pollution test, where all sensors are scattered in the air and transported by the wind; and a vehicle network, where sensor nodes are carried by moving vehicles. Applications are able to collect the data from the sensors about air pollution and traffic conditions. In addition, HDSNs may provide application layer solutions to some existing issues in the network layer. Take network topology adaptivity as an example: when an application observes that the density or the number of sensor nodes in Region X is not sufficient to satisfy the application requirements, it could command the redundant or idle sensor nodes in Region Y to move to Region X.

2.3 Location Awareness

In the context of this paper, we assume that the sensor nodes are location-aware via GPS and other positioning techniques. The location awareness of sensor nodes is very important since sensor networks are deployed and operated in a geographical area after all. Since the sensor nodes in HDSNs are mobile, location information is crucial not only for certain kinds of spatial queries but also for the sensor readings to be meaningful. In addition to the time, sensor ID and readings, location information is frequently used in query predicates and requested by the applications. Moreover, location is frequently used in routing, dissemination and location-based queries [3, 8, 10, 21, 20, 27, 28].

A location needs to be specified explicitly or implicitly for its use. Location models depend heavily on the underlying location identification techniques employed in the system and can be categorized as follows:

- **Geometric Model:** A location is specified as an n-dimensional coordinate (typically $n = 2$ or 3), e.g., the latitude/longitude pair in the *GPS*. The main advantage of this model is its compatibility across heterogeneous systems. However, providing such fine-grained location information may involve considerable cost and complexity.
- **Symbolic Model:** The location space is divided into disjointed zones, each of which is identified by a unique name. Examples are Cricket [18] and the cellular infrastructure. The symbolic model is

in general cheaper to deploy than the geometric model because of the lower cost of the coarser location granularity. Also, being discrete and well-structured, location information based on the symbolic model is easier to manage.

The geometric and symbolic location models have different overheads and levels of precision in representing location information. The appropriate location models to be adopted depends on applications. In this paper, we only consider the geometric location model.

2.4 Window Query

Due to the mobility of sensor nodes, querying the physical world based on IP addresses or IDs of the sensor nodes is not practical. For many applications of sensor networks which need to extract data from a specific geographical area, spatial queries such as window query and nearest neighbor search are essential. In this paper, we focus on window queries.

Window query enables users to retrieve all the data falls within the *query window*, a 2- to 3-dimensional area of interest defined by users. For example, consider a sensor network for an air pollution test, in which all sensors are scattered in the air and transported by the wind. Possible queries are: "What is the *average* pollution index value in a 10-meter space surrounding me?" or "Tell me if the *maximum* air pollution index value in Region X is over α?" In the first query, the query originates from inside the query window, but the latter one is issued from outside the window. In addition, in a vehicle network where sensors are carried by cars. A user may decide to change her driving route dynamically by issuing a query like "How many cars are waiting at the entrance of George Washington Bridge?" As seen in the above examples, practical window queries usually are coupled with aggregation functions, such as *AVG, SUM, MAX*, etc. Thus, aggregation is an important operation to be carried out by the sensor networks. Aggregation algorithms are important not only to provide computational support for those functions but also to reduce the number of messages and energy consumption in the network. How to efficiently aggregate and compute the functions in network is an actively pursued research topic in sensor database. We do not provide specific algorithms for aggregation functions, but focus on issues and strategies in enabling aggregation operation.

2.5 Performance Requirements

In order to assess the various enabling techniques for processing window queries in HDSNs, evaluation criterias need to be considered. In the following, we discuss some performance requirements:

- *Energy efficiency.* Sensor nodes are driven by extremely frugal battery resource, which necessitates the network design and operation be done in an energy-efficient manner. In order to maximize the lifetime of sensor networks, the system needs a suite of aggressive energy optimization techniques, ensuring that energy awareness is incorporated not only into individual sensor node but also into groups of cooperating nodes and the entire sensor network. Based on this remark, our work studies message routing, sensor cooperations, data flow diffusion and aggregation by taking energy efficiency into consideration. These concepts are not simply juxtaposed, but fitting into each other and justify an integrative research topic.
- *Total message volume*: Recent studies show that transmitting and receiving messages dominate the energy consumption on sensor node [19, 23]. Therefore, controlling the total message volume has a significant effect on reducing the energy consumption (in addition to the traffic) of the network. Furthermore, it also reflects the effectiveness of the aggregation and filtering of sensor readings. We expect that aggregation and filtering inside the network can reduce the total message volume tremendously.
- *Access latency*: This metric, indicating the freshness of query results, is measured as the average time between the moment a query is issued and the moment the query result is delivered back to the user. In addition to the lifetime of the sensor networks, access latency is important to the most of applications, especially the ones with critical time constraints. Usually there are tradeoffs between energy consumption and access latency.
- *Result accuracy and precision*: The other performance factors trading off with energy consumption and access latency are result accuracy and precision. High results accuracy and precision requires powerful sensing ability, high sampling rate, localized cooperation among sensor nodes, and larger packet size for transmissions. Approximate results with less precision may sometimes be acceptable by the applications. Network should achieve as high accuracy and precision of query result as other constraints allow.
- Query success rate: Query success rate is the ratio of the number of successfully completed queries against the total number of query issued by applications. This criteria shows how effective the employed query processing algorithms and network protocols are.

3. RESEARCH ISSUES

Although there exist some studies on various related issues of processing window query in highly dynamic sensor networks, they only address some partial aspects of the problems. To the best of our knowledge, this paper presents the first effort to provide a complete suite of solutions/strategies to processing window query in HDSNs. In the following, we investigate the issues by considering the following query execution plan:

1. **Routing a query toward the area specified by its query window.** Once a user (or an application) issues a window query, the first question that needs to be answered is how to bring the query to the targeted area in order to retrieve data from sensor nodes located there. There exist many routing protocols based on state information of the network topology or the neighborhood to a routing node. However, the mobility of sensor nodes in HDSNs makes those protocols infeasible. In HDSNs, the state information changes so frequently that maintaining state consistency represents a major problem. It is very difficult (if not impossible) to obtain a network-wide state in order to route a query efficiently. Thus, *stateless* strategies need to be devised. Here we exploit the location-awareness of sensor nodes to address the need of stateless routing.

   An intuitive stateless routing strategy is *flooding* the network. Since each sensor node is aware of its own location, it can easily decide whether itself is within the query window or not. If a sensor node receives a query and finds itself located within the specified query window, it may return its sensed data back to the sender for processing while re-broadcasting the query to its neighbors. Flooding does not require the sensor nodes to have knowledge of their neighbors and the network in order to route a query to targeted sensor nodes, so it meets the constraints of HDSNs very well. However, all the drawbacks of flooding such as implosion, overlap and resource inefficiency are inherited. In addition, data is very difficult to aggregate by flooding.

   Considering the spatial nature of window queries and the location-awareness of sensor nodes, a class of protocols, called *geo-routing protocols*, that make routing decisions based on locations of sensor nodes and their distances to the destination looks promising. However, most of the existing geo-routing protocols require some knowledge of neighbors' locations to the sensor nodes in order to make a routing decision. In this paper, we propose a stateless geo-routing protocol, called *spatial query routing (SQR)*, which takes the strength of geo-routing protocols and employs various heuristics to fine tune the query routing decisions. Based on SQR, a window query is routed towards the area specified by query window based on sensor nodes' locations and energy awareness, without any state information of neighbors or network topology.

2. **Propagating the query to sensor nodes located within the query window.** Once a query arrives in one of the sensor nodes located in the area specified by the query window, the sensor node may decide whether to start the *query propagation* mode right there or pass the duty to a more suitable sensor node (e.g., send the query to a node located at the center of the window). An algorithm for query propagation should try to satisfy the following two requirements: (1) cover all the sensor nodes located in the window; and (2) terminate query propagation when all the nodes received the query. A strict enforcement of requirement (1) can ensure that no sensor node misses the query. Any miss may lead to an inaccurate query result. However, this requirement is sometimes difficult to satisfy due to the dynamic nature of the sensor networks considered here. Thus, this requirement can be weaken, based on various specifications, to accept an approximated answer or an answer with less precision. Enforcing requirement (2) is critical because the query propagation process should stop once all the sensor nodes in the window receive the query.

   Conventional flooding algorithms can be modified to satisfy the above two requirements. Each sensor node maintain a query cache, which records all the queries it receives. When a sensor node receives a query, it first checks its query cache to see if there is a matching query. If yes, the query will be simply dropped; otherwise, the query is retransmitted to all its neighbors. In this way, query propagation terminates when all the sensor nodes inside the query window have the query in their caches. While the query cache can terminate the query propagation process, overhead inherited from flooding still exists. Furthermore, during the query propagation, sensor nodes may be still in move. Should the new nodes join the query processing? Should the nodes leave the window quit the query? The semantics and implied operations of queries need to be clearly defined.

3. **Collecting and aggregating the data in the query window.** As we pointed out earlier, aggregation is an important operation to be supported in sensor networks for computation of aggregation functions and for reducing the number of message transmissions and energy consumptions in the networks. Thus, instead of having all the sensor nodes located inside a query window send back their readings to the user for further processing, it is more efficient to process the data in network and only deliver the result back. To process

sensor readings based on certain aggregation functions and filtering predicates, the common wisdom is to assign a sensor node located inside the query window as an *aggregation leader*, who collects and processes the readings (or partially computed results) from other nodes. This approach may work for a static network. However, in our scenario, sensor nodes may move constantly so a fixed or static leader may not exist. Therefore, how to locate the leader in order to process the sensor readings locally and correctly is a challenge. In this paper, we introduce a concept of *leading region* to accommodate the mobility of aggregation leaders. Based on spatial space-division, we propose a solution, called *spatial propagation and aggregation* (SPA), for query propagation and data aggregation.

4. **Returning the query result back to the user.** After the query is processed, the results need to be delivered back to the user. Due to the user mobility, delivering the query result back to the user is not trivial. One intuitive approach is to route the result message based on sensor ID. However, in a highly dynamic sensor network, an ID-based routing implies flooding and thus imposes expensive energy and communication overheads. In this paper, we combine the geo-routing and a message forwarding strategy to solve the problem.

4. PROPOSED SOLUTIONS

In this section, we present several innovative solutions that we proposed to address the problems discussed in the previous section. We first describe *SQR*, a stateless spatial query routing method to route a window query towards the area specified by its query window. Then, we present *SPA*, an spatial space-division based approach for query propagation and sensor data aggregation within the query window. Finally, we discuss our solution for returning the query result to a user with mobility.

4.1 Spatial Query Routing

In HDSNs, it is difficult for a *sender*, the sensor node which currently holds a query message, to make query routing decisions without even knowing whether there exists a neighbor. Thus, an idea is to let the potential *query forwarders*, the sensor nodes reachable from the sender, decide whether they would voluntarily forward the query message based on their own state information, such as the distances from the sender and query window, their remaining energy levels, moving directions, speeds, etc. This approach is similar to the implicit geographical forwarding (IGF) protocol proposed in [16]. To facilitate the potential volunteers in making timely and proper decisions, the sender provides information such as

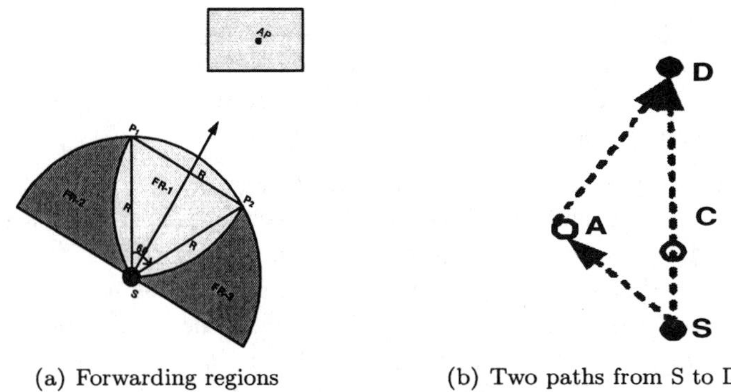

(a) Forwarding regions (b) Two paths from S to D

Figure 1: Spatial query routing.

its own location, the query window (specified by two points), the size of message, and other auxiliary information to prioritize the volunteer query forwarders. SQR consists of two primary tasks: 1) determining a volunteer to serve as the query router; and 2) setting up the next-hop query router based on *overhearing*. The goal of the second task is to reduce forwarding delays.

4.1.1 Volunteer Forwarders

To simplify the presentation, we use Figure 1 to depict a snapshot of the spatial query routing. A sender S, looking for a query forwarder, will broadcast messages to a space of radio range R. The routing of a query message is directed by aiming at a point in the query window, called *Anchor Point* (AP), determined by the application. In Figure 1, the AP is set to be the center of the query window. Here we assume that all the communications are bidirectional.

Once a sensor node receives a query and becomes the sender, it first decides its *forwarding region (FR)* based on the AP and its current position. The forwarding region is the upper part of the circle which is vertical to the line between AP and the sender. The FR is further divided by the sender into three parts:

- **FR-1** is the area with vertices S, P_1 and P_2, and surrounded by three curves. The curve connecting any two vertices is the partial circumstance of the circle centered at the remaining vertex. For example, the curve between P_1 and S is on the circle with center P_2. Therefore, any sensor nodes located inside FR-1 can hear each others' communications with the sender.
- **FR-2 and FR-3** are the two regions inside the FR, but which fall outside of FR-1. In other words, sensor nodes in these regions can communicate with the sender S, but are not necessarily aware of

on-going communications between sender and other sensor nodes in FR. FR-2 and FR-3 are separated by FR-1 (shown as the dark area in Figure 1(a)). Here we number the forward regions based on their forwarding priorities (as explained below).

There are many schemes to assign forwarding priorities to the regions. Here, the sender calculates three priority levels based on the distances of the forwarding regions to the query window. As illustrated in the figure, FR-1 has a higher priority than FR-2 which in turn has a higher priority than FR-3. To find the next query forwarder, the sender broadcasts a *Forwarder_Request* message to all its neighbors. The message contains the current location of the sender, query window, location of AP, and three timer (i.e., *Response_timer*$_i$, where $1 \leq i \leq 3$) for the three FRs described above. Response_timer$_i$ specifies the allocated period of time sensor nodes located in FR-i should respond to the sender. The responding priorities of sensor nodes in the same FR are determined by parameters such as its energy level, distance to sender and destination, etc.

Once receiving the message, a potential forwarder checks whether itself is located inside a FR. If not, the sensor node simply drops the message. Otherwise, based on which FR it is located within and the parameters considered below, the node computes and holds itself for a certain period of *holding time*. During the holding time, if the node does not hear from other neighbor nodes or the sender, it sends an *ACK_Forwarder* message back to the sender after the timer has expired; otherwise, it drops the message and cancels the holding timer. A node N inside FR may use some heuristics based on the following parameters to determine the holding time:

- **Remaining energy on the node** N. It aims to balance the energy consumption in the network. The node with higher energy remaining has a shorter holding time, thus suppresses the acknowledgements from other nodes in the same FR with less energy resources.
- **The distance between** S **and** N. The longer the distance between S and N, the shorter the holding time. The metric tries to reduce the number of hops between the source and the destination by forwarding the message as far away from S as possible. However to calculate the distance, S has to attach its location with the message. Moreover, using this metric may cause a longer path from the source to the destination. For example in Figure 1(b), S and D represent the source and the destination respectively. There are two possible paths from S to D: either $S \rightarrow C \rightarrow D$ or $S \rightarrow A \rightarrow D$. Based on this metric, the packet will follow the path of $S \rightarrow A \rightarrow D$.

Obviously it is not the best path from S to D, because longer transmission distance costs more energy.

- **The distance between N and the query window**. To overcome the side effect in the previous metric, the nodes can also take account into the distance between itself and the query window[1]. The N who has longer distance from the distance has longer timer.
- **Forwarding regions**. The sensor nodes located in FR-1 have a higher priority than those located in FR-2 and FR-3 to serve as a forwarder. The sensor nodes in FR-1 have to respond to the sender before $Response\_timer_1$ has expired. The sensor nodes in FR-2 will need to respond after $Response\_timer_1$ and before $Response\_timer_2$ has expired. The sensor nodes in FR-3 follow the same rule. Thus, nodes in FR-1 have shorter holding timers than the ones in FR-2, which have shorter timer than the ones in FR-3.

The situation where the the sender does not receive any *Ack_Forwarder* message from any of FRs implies that there is no node in FRs of the sender or that there exist some nodes there but none of them are willing to serve as a forwarder. This well-known *forwarding hole problem* can be resolved by existing solutions [10] or by re-try after waiting for a short period of time (since the network is dynamic).

4.1.2 Pipelined Forwarding

In the forwarder volunteering process described above, the query forwarder has to wait for quite a while (i.e., after sending an *ACK_Forwarder* to the sender and receiving the query message from the sender) before it can look for the next forwarder. In this section, we introduce a pipelined forwarding strategy based on overhearing to reduce forwarding delay.

When a forwarder acknowledges the sender by sending an *ACK_Forwarder* message, the nodes inside forwarder's FRs also can overhear the message. To speed up the query forwarding process, we propose to have a potential forwarder to attach query forwarding information (which was given in *Forwarder_Request* message) in the *ACK_Forrwarder* message. Thus, the *ACK_Forrwarder* message is not only serving as the acknowledgement to the sender but also used to find the next forwarder.

For example, in Figure 2(a), when S hears an ACK from the forwarder, the sensor nodes in FRs circled by the dotted line can also hear the ACK, which contains enough information for those sensor nodes to decide whether to volunteer as the next forwarder. Figure 2(b) shows the timeline for three transmissions. Here is a simple analysis of the idea (assuming

[1] Depending on derived heuristics, the distance can be between N and the center, the nearest points, or AP of the window.

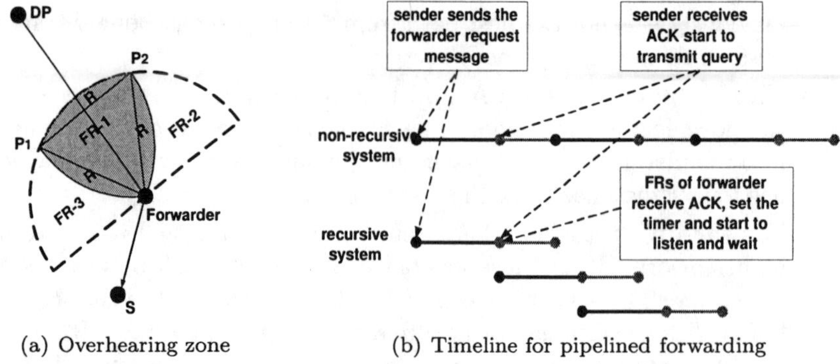

(a) Overhearing zone

(b) Timeline for pipelined forwarding

Figure 2: Pipelined forwarding.

no conflicts happen). If the average *Response_Timer* is RT seconds, the average query transmission time is QT seconds, and the query message is forwarded M times before arriving the DP, the non-pipelined system takes $M \times (RT + QT)$ seconds for query forwarding, but the pipelining system only needs $M \times (RT + QT) - (M - 1) \times QT = M \times RT + QT$, and saved $(M - 1) \times QT$ seconds. We feel this idea is feasible and we are currently developing more detailed analysis to validate it.

4.2 Spatial Propagation and Aggregation

HDSNs bring major challenges to query propagation and data aggregation. Due to the mobility of sensor nodes and frequently changing network topology, the static aggregation leader and fixed infrastructure that is based on existing query propagation and data collection algorithms may not exist [14, 25, 28]. In this section, we propose a query propagation and data collection algorithm, called *spatial propagation and aggregation* (SPA), based on spatial space division and the concept of *leading regions* (LR), which are designated regions for dispatching query and processing aggregation. One of the sensor nodes in the leading region will serve as a *propagation and aggregation leader* (PAL) to decide how to propagate a query to space-divided subareas and to collect data from those subareas for aggregation. The size and location of a leading region can be present or adapted during run-time. The PAL of a leading region can be determined by volunteering (based on various constraints of the sensor nodes), voting (based on group decision), or designating (e.g., the first node receives the query). In addition, the PAL can adjust its LR before it proceeds to propagate queries. The size of LR can be determined by the mobility of PAL and density of nodes in the area. The main difference between the leading area here and the leader in static sensor networks is that a leading area addresses a fixed geographical area instead of some

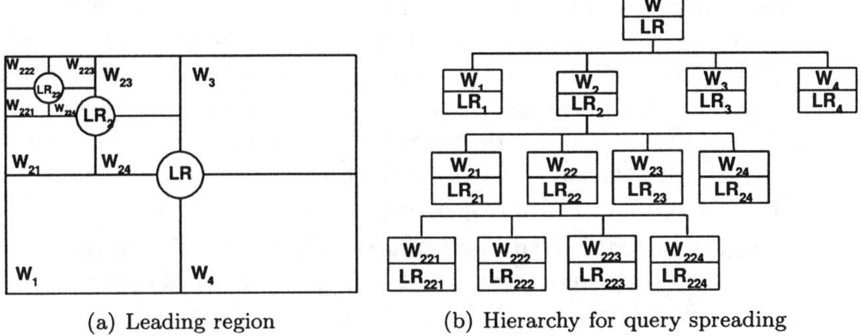

(a) Leading region (b) Hierarchy for query spreading

Figure 3: Spatial propagation and aggregation.

specific nodes which are dynamic in HDSNs.

Based on the ideas discussed above, we use Figure 3 to further illustrate our SPA algorithm. While the leading areas can be of any shape, we represented them as circles to distinguish from the sub-windows. Once a query is received by a sensor node in the query window, the sensor node can decide to 1) serve as the PAL for the window by specifying an LR and start the SPA algorithm; or 2) forward the query to a temporary LR (e.g., center of the window) and allow more qualified nodes there to volunteer. Assume that a sensor node in the temporary LR decides to serve as the PAL, it will specify its LR and then start the SPA algorithm.

The SPA algorithm we proposed is recursive. Once the PAL in charge of the whole window and its LR are determined, the query window W is divided into four sub-windows, namely W_1, W_2, W_3 and W_4. For each sub-window W_i where i is the subscript of the sub-window, a circle area in the middle is temporarily assigned as the leading region, denoted as LR_i. The query will be propagated to the four leading regions by using SQR protocol described earlier. After a query arrives at the temporary leading region in the sub-window, a PAL in charge of the sub-window is determined and its refined LR are obtained. Thus, each sub-window (W_i) is further divided into even smaller windows, i.e., W_{i1}, W_{i2}, W_{i3} and W_{i4}. The recursive splitting and forwarding procedure is repeated until a stop condition is satisfied, which will be discussed later in this section. Once the splitting and forwarding process stops, sensor nodes report their readings back to their leading region where the PAL will collect and aggregate the data. Once the PAL decides that there are no more reports from the area it leads, it sends the aggregated data to the leading region at an immediate higher level. This aggregation process is repeated until the PAL for the whole query window receives the aggregation result from all of its children LRs. However, there are still several questions need to be answered in the above seemingly simple recursive algorithm:

1. **How large should the leading region be?** In SPA, the leading region is set by the PAL based on its mobility and the network density, so that there is always at least one sensor node in the region. Further analysis is needed to obtain the optimal leading region. To address the rare scenario that there is no sensor in the leading region, one possible solution is to extend the routing protocol such that the data may be sent to the upper-level leading areas.
2. **How do we deal with the scenario that the PAL moves out of the leading region before the aggregation is completed?** Once the PAL realizes that it is close to the boundary of leading region and probably will move out of the region, it floods the leading region to find a new PAL. When a new PAL is found, the old PAL hands off the partially completed aggregation result and other needed information (e.g., the higher level leading region, information about the lower level region) to the new PAL. As such, the data collection and aggregation process is not interrupted.
3. **What is the stop condition for recursive splitting and forwarding?** Many heuristics can be used. For example, the size of sub-windows or the ratio of window size versus LR size. Once a PAL stops window splitting and query propagation, it floods the query inside the sub-window that it is in charge and waits for the data from the sensor nodes inside its window. Since flooding is constrained in a very small region, duplicated transmissions are reduced.

4.3 Returning the Query Result Back to the (Mobile) User

Once the PAL at the highest level obtains the query result, it will return the result back to the user (i.e., the query issuer). Due to the unlimited mobility, a user may move while the query is processed. Therefore the location where the query was issued may not be the exact location where the result should be delivered.

To address this problem, an *issuer region*, which defines the region where the user may stay in (while moving) during query processing, will be included as part of the query routed towards the query window. Thus, instead of routing the query result based on the user's identifier or exact location, the stateless SQR protocol (as described in Section 4.1) is used to send the result back to the issuer region. Once the query result reaches the issuer region, it is flooded to the whole region.

Similar to decide a leading region in the SPA algorithm, a user should specify a precise area that will accommodate his/her future movement. However, in the rare occasion that the user moves out of the issuer region, it will specify a new region which may or may not overlap with the original region, called the *extended issuer region* (EIR), and pass this information to a sensor node, called the *forwarding agent* (FA), located in the old region. If the FA has to move out of the old issuer region, it will find a substitute to serve the role of forwarding agent and pass the information of EIR to the new FA. This forwarding mechanism will be established if the user has to move out of the new EIRs too. Thus, the query result may go across several forwarding agents until reaching the user.

5. RELATED WORK

To the best of our knowledge, window query process in highly dynamic sensor networks has not yet been studied in the sensor network and database research community. However, our work has been informed and influenced by a number of research efforts.

Energy-efficient data dissemination is among the first set of related research issues being addressed. SPIN is one of the early works that is designed to address the deficiencies of classic flooding by negotiations and resource adaption [5, 12]. Directed Diffusion takes a data-centric naming approach to enable in-network data aggregation [9]. By employing initial low-rate data flooding and gradual reinforcement of better paths, directed diffusion accommodates a certain level of sensor nodes dynamics. TTDD proposed by Ye et.al targets a scalable and efficient data delivery to multiple mobile sinks in a large sensor network [27]. TTDD enables mobile sinks to continuously receive data on the move at the cost of building up a grid structure for each data source within the whole network. Instead of purely flooding data or a request, rumor routing algorithm proposed by Braginsky et al. is a logical compromise [2]. A source sends out"agents" which randomly walk in the sensor network to set up event paths. Queries also randomly walk in the sensor field until they meet an event path. Such previous work investigates energy-efficient data dissemination protocols in stationary or low-mobility sensor networks, since frequent movements of upstream nodes on the route may cause a new route discovery procedure or high route maintenance expense. In our proposal, sensor nodes make decisions to route a query without any state information of other nodes or networks, thus suitable for the sensor networks with unstable topology.

In addition, recent work has explored the routing algorithm for window query in static sensor networks [28]. Specifically GEAR showed how the geographical forwarding technique can help achieve energy efficiency for routing the query toward the window and spreading the query inside the window. Our ideas not only explore the query forwarding and dissemination in a stateless way, but also the data collection and aggregation and delivering the result back to the users in HDSN.

Some of the inspiration for our studies comes from [26], which enables the distributed and localized sensor operations to improve the scalability. Ye et.al proposed an original forwarding scheme for maintaining relatively constant working node density, thus prolonging the system lifetime.

Many of the energy-aware techniques developed for improving sensor network performance can be adopted by our study. For example LEACH [4] proposes a clustering based protocol that utilizes randomized rotation of local cluster heads to evenly distribute the energy load among the sensors in the network. Our proposal has the additional degree of freedom in being able to use application semantics to achieve further efficiency.

Our work borrows heavily from the literature on ad hoc routing. Specifically, it is close related to the geographical routing algorithm proposed in [10] where each sensor node greedily forwards the packet based on the knowledge of its neighbors. As a state-based algorithm, it suffers expensive upkeep of neighbor changes due to the system dynamics. The LAR, in its attempt to constrain the flooding into smaller possible region instead of the whole network, is another closely related work [11]. However the effectiveness of the estimation for requested zone in the LAR is restricted by the destination's previous known location and its know mobility pattern.

Implicit geographical forwarding (IGF) proposed by B.M. Blum et. al [16] is probably the most relevant work to the SQR discussed in the paper. IGF designs a state-free routing algorithm for sensor networks, by assuming high node density and location awareness in the networks. In IGF, sensor nodes make forwarding decisions dynamically based on their current conditions (i.e., distance to the destination and remaining energy). However, this work has focused on adapting the underlying MAC layer to avoid the communication collisions. Moreover, IGF considers only the sensor nodes residing in a small region (which is part of the FR-1 in SQR) for forwarding and relies on shifting the forwarding area to handle the situation when there are no sensor nodes in the forwarding zone. Our SQR protocol tries a much larger area for forwarding and sets up different forwarding zones. The issue we are focused on is how to set up the minimal zone timers while reducing collisions within a zone and to ensure no communication interference between the zones. Specifically, SQR defines much wider forwarding region (which consists of FR-1, FR-2 and FR-3) and assigns them different forwarding priorities. Another primary difference between IGF and SQR is that pipelining is employed by SQR to reduce the forwarding latency. Therefore, we are interested in exploiting (via SQR) some of the related issues considered in [16] but in a more sophisticated context.

Our work is influenced by the data management researches in sensor networks as well. In Moving Objects Database (MOD), a base station (gateway) outside the network collects the sensor readings and answers user queries [22, 24]. However the centralized approach incurs heavy transmission overhead and consumes extensive energy. Our work focuses on extracting and aggregating the data locally inside the network, thereby reducing the duplicate and unnecessary transmissions. In both TinyDB and Cougar projects, the query is executed directly on the sensor nodes instead of on the gateway. Therefore, the sensor readings are only extracted out when a query is inserted into the network. TinyDB develops the techniques for query result aggregation and filtering on site by building up a semantic tree [13, 14]. Cougar project studies the query execution plan and query optimization in the sensor networks [25].

6. CONCLUSION

Motivated by future sensor network applications, we studied the problem of window query execution in highly dynamic sensor networks, where the network topology changes frequently. The proposed suite of protocols forwards the query toward the queried window based on the location of the sensor nodes; propagates the query inside the query window recursively and ensures the coverage of the window; aggregates the sensor readings at certain geographical region before delivering it back (which adapts to the mobility of sensor nodes and reduces the amount of transmissions); guarantees that the query issuer receives the result even if he is in movement. Our protocols explore the static property of geographical location to accommodate the mobility of both sensor nodes and users.

Some issues remain to be further exploited such as the heuristics for forwarder selection, zone timer settings, and the size of the leading regions. Other future work includes developing simulations and implement prototypes to validate our proposals.

7. REFERENCES

[1] G. Asada, T. Dong, F. Lin, G. Pottie, W. Kaiser, and H. Marcy, 1998. Wireless integrated network sensors: Low power systems on a chip. In *Proc. European Solid State Circuits Conference*, Hague, Netherlands.

[2] D. Braginsky and D. Estrin, September 2002. Rumor routing algorithm for sensor networks. In *Proceedings of the First ACM International Workshop on Wireless Sensor Networks and Applications*, pages 22–31, Atlanta, GA.

[3] J. Heidemann, F. Silva, C. Intanagonwiwat, R. Govindan, D. Estrin, and D. Ganesan, October 2001. Building efficient wireless sensor

networks with low-level naming. In *Proceedings of the Symposium on Operating Systems Principles*, pages 146–159, Banff, Alberta, Canada.

[4] W. R. Heinzelman, A. Chandrakasan, and H. Balakrishnan, January 2000. Energy-efficient communication protocol for wireless microsensor networks. In *IEEE Proceedings of the Hawaii International Conference on System Sciences (HICSS)*, Maui, Hawaii.

[5] W. R. Heinzelman, J. Kulik, and H. Balakrishnan, 1999. Adaptive protocols for information dissemination in wireless sensor networks. In *Proceedings of the 5th Annual ACM/IEEE International Conference on Mobile Computing and Networking*, pages 174–185, Seattle, WA.

[6] J. Hightower and G. Borriello, August 2001. Location systems for ubiquitous computing. *IEEE Computer*, 34(8):57–66.

[7] J. Hill, R. Szewczyk, A. Woo, S. Hollar, D. Culler, and K. Pister, 2000. System archtecture directions for networked sensors. *ACM SIGPLAN Notices*, 35(11):93–104.

[8] T. Imielinski and S. Goel, October 2000. Dataspace - querying and monitoring deeply networked collections in physical space. *IEEE Personal Communications*, 7(5):4–9.

[9] C. Intanagonwiwat, R. Govindan, and D. Estrin, 2000. Directed diffusion: a scalable and robust communication paradigm for sensor networks. In *Proceedings of the 6th Annual International Conference on Mobile Computing and Networking*, pages 56–67, Boston, MA.

[10] B. Karp and H. T. Kung, 2000. GPSR: greedy perimeter stateless routing for wireless networks. In *Proceedings of the 6th Annual International Conference on Mobile Computing and Networking*, pages 243–254, Boston, MA.

[11] Y.-B. Ko and N. H. Vaidya, 2000. Location-aided routing (LAR) in mobile ad hoc networks. *Wireless Networks*, 6(4):307–321.

[12] J. Kulik, W. R. Heinzelman, and H. Balakrishnan, 2002. Negotiation-based protocols for disseminating information in wireless sensor networks. *Wireless Networks*, 8(2-3):169–185.

[13] S. Madden and M. J. Franklin, February 2002. Fjording the stream: an architecture for queries over streaming sensor data. In *18th International Conference on Data Engineering (ICDE)*, San Jose, CA.

[14] S. Madden, M. J. Franklin, J. M. Hellerstein, and W. Hong, 2002. TAG: a tiny aggregation service for ad-hoc sensor networks. *ACM SIGOPS Operating Systems Review*, 36(SI):131–146.

[15] A. Mainwaring, D. Culler, J. Polastre, R. Szewczyk, and J. Anderson, 2002. Wireless sensor networks for habitat monitoring. In *Proceedings of the 1st ACM International Workshop on Wireless Sensor Networks and Applications*, pages 88–97, Atlanta, GA.

[16] B. M.Blum, T. He, S. Son, and J. A. Stankovic, 2003. IGF: A robust state-free communication protocol for sensor networks. Technical Report CS-2003-11, University of Virginia.

[17] N. Patwari, A. III, M. Perkins, N. Correal, and R. O'Dea, August 2002. Relative location estimation in wireless sensor networks. *IEEE Transactions on Signal Processing, Special Issue on Signal Processing in Networks*, 51(9):2137–2148.

[18] N. B. Priyantha, A. Chakraborty, and H. Balakrishnan, 2000. The cricket location-support system. In *Proceedings of the 6th Annual International Conference on Mobile Computing and Networking*, pages 32–43, Boston, MA.

[19] V. Raghunathan, C. Schurgers, S. Park, and M. B. Srivastava, March 2002. Energy aware wireless microsensor networks. *IEEE Signal Processing Magazine*, 19(2):40–50.

[20] S. Ratnasamy, B. Karp, S. Shenker, D. Estrin, R. Govindan, L. Yin, and F. Yu, 2003. Data-centric storage in sensornets with ght, a geographic hash table. *Mobile Networks and Applications*, 8(4):427–442.

[21] S. Ratnasamy, B. Karp, L. Yin, F. Yu, D. Estrin, R. Govindan, and S. Shenker, September 2002. GHT: a geographic hash table for data-centric storage. In *First ACM International Workshop on Wireless Sensor Networks and Applications (WSNA)*, Atlanta, GA.

[22] A. P. Sistla, O. Wolfson, S. Chamberlain, and S. Dao, April 1997. Modeling and querying moving objects. In *13th International Conference on Data Engineering (ICDE)*, pages 422–432, Birmingham, U.K.

[23] WINS project. Electrical engineering department, ucla. http://www.janet.ucla.edu/WINS/.

[24] O. Wolfson, B. Xu, S. Chamberlain, and L. Jiang, April 1998. Moving objects databases: issues and solutions. In *Statistical and Scientific Database Management*, pages 111–122, Capri, Italy.

[25] Y. Yao and J. Gehrke, January 2003. Query processing for sensor networks. In *Proceedings of the First Biennial Conference on Innovative Data Systems Research*, pages 21–32, Asilomar, CA.

[26] F. Ye, S. Lu, and L. Zhang, April 2001. GRAdient broadcast: a robust, long-lived large sensor network. Technical report, http://irl.cs.ucla.edu/papers/grab-tech-report.ps.

[27] F. Ye, H. Luo, J. Cheng, S. Lu, and L. Zhang, 2002. A two-tier data dissemination model for large-scale wireless sensor networks. In *Proceedings of the Eighth Annual International Conference on Mobile Computing and Networking*, pages 148–159, Atlanta, GA.

[28] Y. Yu, R. Govindan, and D. Estrin, May 2001. Geographical and energy aware routing: A recursive data dissemination protocol for wireless sensor networks. Technical Report UCLA/CSD-TR-01-0023, UCLA Computer Science Department.

[29] F. Zhao, J. Shin, and J. Reich, March 2002. Information-driven dynamic sensor collaboration for tracking applications. *IEEE Signal Processing Magazine*, 19(2):61–72.

Approximate Query Answering on Sensor Network Data Streams

Alfredo Cuzzocrea[2], Filippo Furfaro[2], Elio Masciari[1], Domenico Saccà[1], and Cristina Sirangelo[2]

[1] ICAR-CNR – Institute of Italian National Research Council
{masciari, sacca}@icar.cnr.it
[2] DEIS-UNICAL
Via P. Bucci, 87036 Rende (CS) Italy
{cuzzocrea, furfaro, sirangelo}@si.deis.unical.it

ABSTRACT

Sensor networks represent a non traditional source of information, as readings generated by sensors flow continuously, leading to an infinite stream of data. Traditional DBMSs, which are based on an exact and detailed representation of information, are not suitable in this context, as all the information carried by a data stream cannot be stored within a bounded storage space. Thus, compressing data (by possibly loosing less relevant information) and storing their compressed representation, rather than the original one, becomes mandatory. This approach aims to store as much information carried by the stream as possible, but makes it unfeasible to provide exact answers to queries on the stream content. However, exact answers to queries are often not necessary, as approximate ones usually suffice to get useful reports on the world monitored by the sensors. In this paper we propose a technique for providing fast approximate answers to aggregate queries on sensor data streams. Our proposal is based on a hierarchical summarization of the data stream embedded into a flexible indexing structure, which permits us to both access and update compressed data efficiently. The compressed representation of data is updated continuously, as new sensor readings arrive. When the available storage space is not enough to store new data, some space is released by compressing the "oldest" stored data progressively, so that recent information (which is usually the most relevant to retrieve) is represented with more detail than old one.

1. INTRODUCTION

Sensors are non-reactive elements which are used to monitor real life phenomena, such as live weather conditions, network traffic, etc. They are usually organized into networks where their readings are transmitted using low level protocols [9]. Sensor networks represent a non traditional source of information, as readings generated by sensors flow continuously, leading to an infinite stream of data. Traditional DBMSs, which are based on a detailed representation of information, are not suitable in this context, as all the information carried by a data stream cannot be stored within a bounded storage space [2–4, 7, 8]. Moreover query answering in traditional DBMSs is based on an "exact" paradigm, that is answers are evaluated exactly by accessing at least all the data involved in the query. This can lead to unacceptable inefficiency when the query is issued on a huge amount of data, which is very common for queries which extract summary information (using aggregate operators such as sum, mean, count, etc.) for analysis purposes. The issue of defining new query evaluation paradigms to provide fast answers to aggregate queries is very relevant in the context of sensor networks. In fact, the amount of data produced by sensors is very large and grows continuously, and the queries need to be evaluated very quickly, in order to make it possible to perform a timely "reaction to the world" . Moreover, in order to make the information produced by sensors useful, it should be possible to retrieve an up-to-date "snapshot" of the monitored world continuously, as time passes and new readings are collected. For instance, a climate disaster prevention system would benefit from the availability of continuous information on atmospheric conditions in the last hour. If the answer to these queries, called *continuous queries*, is not fast enough, we could observe an increasing delay between the query answer and the arrival of new data, and thus not a timely reaction to the world. In this paper we propose a technique for providing fast approximate answers to aggregate queries on sensor data streams. Our proposal is based on a hierarchical summarization of the data stream embedded into a flexible indexing structure, which permits us to both access and update compressed data efficiently. The compressed representation of data is updated continuously, as new sensor readings arrive. When the available storage space is not enough to store new data, some space is released by compressing the "oldest" stored data progressively, so that recent information (which is usually the most relevant to retrieve) is represented with more detail than old one. Consider, as an example, a network congestion detection system that has to prevent network failures exploiting the knowledge of network traffic during time. To avoid a crash of the network, the system needs to locate the nodes where the amount of traffic has increased in an abnormal way in the last minutes. Thus, the knowledge of the traffic level in the network during the last minutes is more significant for the system than that of the traffic occurred in the last days.

2. PROBLEM STATEMENT

Consider an ordered set of n sources (i.e. sensors) denoted by $\{s_1, \ldots, s_n\}$ producing n independent streams of data, representing sensor readings. Each data stream can be viewed as a sequence of triplets $\langle id_s, v, ts \rangle$, where: 1) $id_s \in \{1, .., n\}$ is the source identifier; 2) v is a non negative integer value representing the measure produced by the source identified by id_s; 3) ts is a *timestamp*, i.e. a value that indicates the time when the reading v was produced by the source id_s.

The data streams produced by the sources are caught by a *Sensor Data Stream Management System* (SDSMS), which combines the sensor readings into a unique data stream, and supports data analysis.

An important issue in managing sensor data streams is aggregating the values produced by a subset of sources within a time interval. More formally, this means answering a *range query* on the overall stream of data generated by $s_1, \ldots, s_n$. A range query is a pair $Q = \langle s_i..s_j, [t_{start}..t_{end}] \rangle$ whose answer is the evaluation of an aggregate operator (such as *sum*, *count*, *avg*, etc.) on the values produced by the sources $s_i, s_{i+1}, \ldots, s_j$ within the time interval $[t_{start}..t_{end}]$.

We point out that considering the set of sources as an ordered set implies the assumption that the sensors in the network can be organized according to a linear ordering. Whenever any implicit linear order among sources cannot be found (for instance, consider the case that sources are identified by a geographical location), a mapping should be defined between the set of sources and a one-dimensional ordering. This mapping should be closeness-preserving, that is sensors which are "close" in the network should be close in the linear ordering. Obviously, it is not always possible to define a liner ordering such that no information about the "relative" location of every source w.r.t. each other is lost. It can happen that two sources which can be considered as contiguous in the network are not located in contiguous positions according to the linear ordering criterion. In this case, a range query involving a set of contiguous sensors in the network is possibly translated into more than one range query on the linear paradigm used to represent the whole set of sources.

The sensor data stream can be represented by means of a two-dimensional array, where the first dimension corresponds to the set of sources, and the other one corresponds to time. In particular, the time is divided into intervals Δt_j of the same size. Each element $\langle s_i, \Delta t_j \rangle$ of the array is the sum of all the values generated by the source s_i whose timestamp is within the time interval Δt_j. Obviously the use of a time granularity generates a loss of information, as readings of a sensor belonging to the same time interval are aggregated. Indeed, if a time granularity which is appropriate for the particular context monitored by sensors is chosen, the loss of information will be negligible.

Using this representation, an estimate of the answer to a sum range query over $\langle s_i..s_j, [t_{start}..t_{end}]\rangle$ can be obtained by summing two contributions. The first one is given by the sum of those elements which are completely contained inside the range of the query (i.e. the elements $\langle s_k, \Delta t_l\rangle$ such that $i \leq k \leq j$ and Δt_l is completely contained into $[t_{start}..t_{end}]$). The second one is given by those elements which partially overlap the range of the query (i.e. the elements $\langle s_k, \Delta t_l\rangle$ such that $i \leq k \leq j$ and $t_{start} \in \Delta t_l$ or $t_{end} \in \Delta t_l$). The first of these two contributions does not introduce any approximation, whereas the second one is generally approximate, as the use of the time granularity makes it unfeasible to retrieve the exact distribution of values generated by each sensor within the same interval Δt_l. The latter contribution can be evaluated by performing linear interpolation, i.e. assuming that the data distribution inside each interval Δt_i is uniform (*Continuous Values Assumption - CVA*). For instance, the contribution of the element $\langle s_2, \Delta t_3\rangle$ to the sum query represented in Fig. 1 is given by $\frac{6-5}{2} \cdot 4 = 2$. As the stream of readings produced by every source is

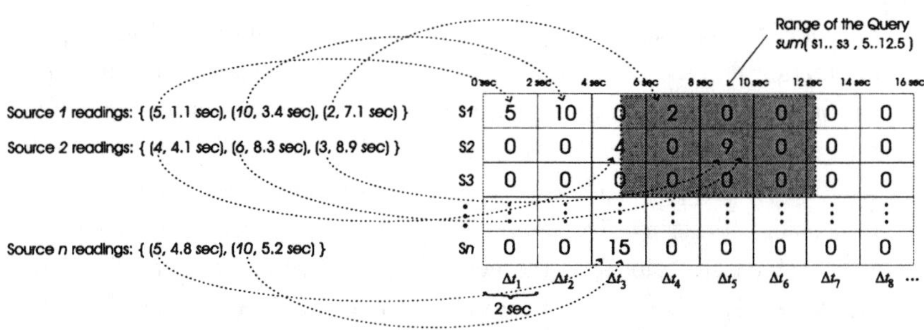

Fig. 1. Two-dimensional representation of sensor data streams.

potentially "infinite", detailed information on the stream (i.e. the exact sequence of values generated by every sensor) cannot be stored, so that exact answers to every possible range query cannot be provided. However, exact answers to aggregate queries are often not necessary, as approximate answers usually suffice to get useful reports on the content of data streams, and to provide a meaningful description of the world monitored by sensors.

A solution for providing approximate answers to aggregate queries is to store a compressed representation of the overall data stream, and then to run queries on the compressed data. The use of a time granularity introduces a form of compression, but it does not suffice to represent the whole stream of data, as the stream length is possibly infinite. An effective structure for storing the information carried by the data stream should have the following characteristics: i) it should be efficient to update, in order to catch the continuous stream of data coming from the sources; ii) it should provide an up-to-date representation of the sensor readings, where recent information is possibly represented more accurately than old one; iii) it should permit us to answer range queries efficiently.

Our proposal. In this paper we propose a technique for providing (fast) approximate answers to aggregate queries on sensor data streams, focusing our attention on *sum* range queries. Our proposal consists in a compressed representation of the sensor data stream where the information is summarized in a hierarchical fashion. In particular, a flexible indexing structure is embedded into the compressed data, so that information can be both accessed and updated efficiently. In more detail, our compression technique works as follows.

- the sensor data stream is divided into "time windows" of the same size: each window consists of a finite number of contiguous unitary time intervals Δt_i (the size of each Δt_i corresponds to the granularity);
- time windows are indexed, so that windows involved in a range query can be accessed efficiently;
- as new data arrive, if the available storage space is not enough for their representation, "old" windows are compressed (or possibly removed) to release the storage space needed to represent new readings, and the index is updated to take into account the new data.

The technique used for compressing time windows is *lossy*, so that "recent" data are generally represented more accurately than "old" data. In Fig. 2, the partitioning scheme of a stream into time windows is represented, as well as the overlying index referring to all the time windows.

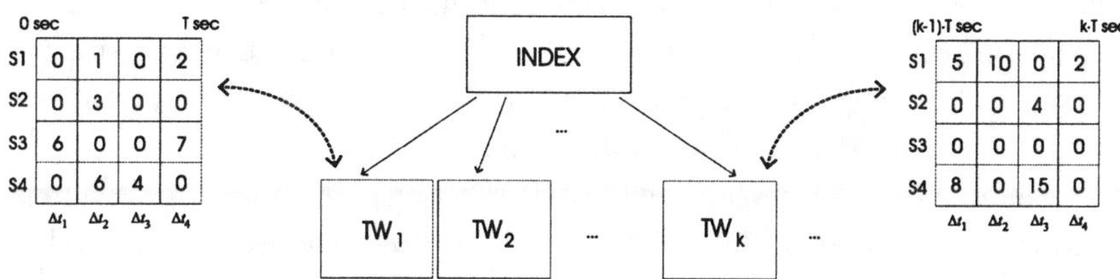

Fig. 2. A sequence of indexed time windows

3. REPRESENTING TIME WINDOWS

3.1 Preliminary Definitions

Consider given a two-dimensional $n_1 \times n_2$ array A. Without loss of generality, array indices are assumed to range respectively in $1..n_1$ and $1..n_2$. A *block* r (of the array) is a two dimensional interval $[l_1..u_1, l_2..u_2]$ such that $1 \leq l_1 \leq u_1 \leq n_1$ and $1 \leq l_2 \leq u_2 \leq n_2$. Informally, a block represents a "rectangular" region of the array. We denote by $size(r)$ the size of the block r, i.e. the value $(u1 - l_1 + 1) \cdot (u_2 - l_2 + 1)$. Given a pair $\langle v_1, v_2 \rangle$ we say that $\langle v_1, v_2 \rangle$ is inside r if $v_1 \in [l_1..u_1]$ and $v_2 \in [l_2..u_2]$. We denote by $sum(r)$ the sum of the array elements occurring in r, i.e. $sum(r) = \sum_{\langle i,j \rangle inside\ r} A[i,j]$. If r is a block corresponding to the whole array (i.e. $r = [1..n_1, 1..n_2]$), $sum(r)$ is also denoted by $sum(A)$. A block r such that $sum(r) = 0$ is called a *null block*. Given a block $r = [l_1..u_1, l_2..u_2]$ in A, we denote by r_i the $i-$th quadrant of r, i.e. $r_1 = [l_1..m_1, l_2..m_2], r_2 = [m_1+1..u_1, l_2..m_2], r_3 = [l_1..m_1, m_2+1..u_2]$, and $r_4 = [m_1 + 1..u_1, m_2 + 1..u_2]$. where $m_1 = \lfloor (l_1 + u_1)/2 \rfloor$ and $m_2 = \lfloor (l_2 + u_2)/2 \rfloor$. Given a a time interval $t = [t_{start}..t_{end}]$ we denote by $size(t)$ the size of the time interval t, i.e. $size(t) = t_{end} - t_{start}$. Furthermore we denote by $t_{i/2}$ the i-th half of t. That is $t_{1/2} = [t_{start}..(t_{start} + t_{end})/2]$ and $t_{2/2} = [(t_{start} + t_{end})/2..t_{end}]$. Given a tree T, we denote by $Root(T)$ the root node of T and, if p is a non leaf node, we denote the $i-$th child node of p by $Child(p, i)$. Given a triplet $x = \langle id_s, v, ts \rangle$, representing a value generated by a source, id_s is denoted by $id_s(x)$, v by $value(x)$ and ts by $ts(x)$.

3.2 The Quad-Tree Window

In order to represent data occurring in a time window, we do not store directly the corresponding two-dimensional array, indeed we choose a hierarchical data structure, called *quad-tree window*, which offers some advantages: it makes answering (portions of) range queries internal to the time window more efficient to perform (w.r.t. a "flat" array representation), and it stores data in a straight compressible format, that is, data is organized according to a scheme that can be directly exploited to perform compression.

This hierarchical data organization consists in storing multiple aggregations performed over the time window array according to a quad-tree partition. This means that we store the sum of the values contained in the whole array, as well as the sum of the values contained in each quarter of the array, in each sixteenth of the array and so on, until the single elements of the array are stored. Fig. 3 shows an example of quad-tree partition, where each node of the quad-tree is associated with the sum of the values contained in the corresponding portion of the array.

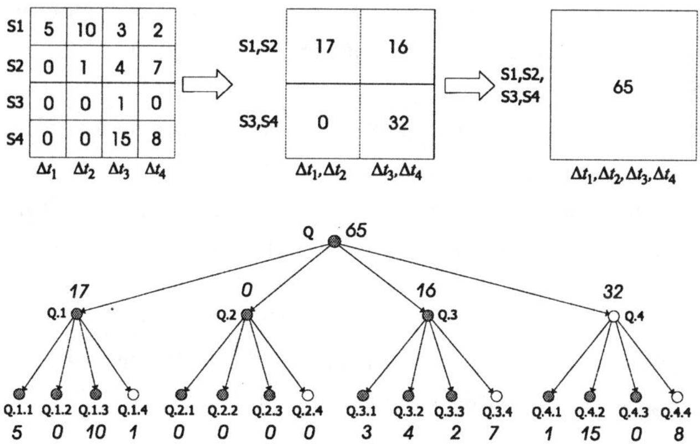

Fig. 3. A Time Window and the corresponding quad-tree partition

The quad-tree structure is very effective for answering (sum) range queries inside a time window efficiently, as we can generally use the pre-aggregated sum values in the quad-tree nodes for evaluating the answer (see Section 6.1 for more details). Moreover, the space needed for storing the quad-tree representation of a time window is about the same as the space needed for a flat representation, as we will explain later. Furthermore, the quad-tree structure is particularly prone to progressive compressions. In fact, the information represented in each node is summarized in its ancestor nodes. For instance, the node Q of the quad-tree in Fig. 3 contains the sum of its children $Q.1$, $Q.2$, $Q.3$, $Q.4$; analogously, $Q.1$ is associated to the sum of $Q1.1$, $Q1.2$, $Q1.3$, $Q1.4$, and so on. Therefore, if we prune some nodes from the quad-tree, we do not lose every information about the corresponding portions of the time window array, but we represent them with less accuracy. For instance, if we removed the nodes $Q1.1, Q1.2, Q1.3, Q1.4$, then the detailed values of the readings produced by the sensors S_1 and S_2 during the time intervals Δt_1 and Δt_2 would be lost, but it would be kept summarized in the node $Q.1$. The compression paradigm that we use for quad-tree windows will be better explained in Section 5.

We will next describe the quad-tree based data representation of a time window formally. Denoting by u the time granularity (i.e. the width of each interval Δt_j), let $T = n \cdot u$ be the time window width (where n is the number of sources). We refer to a *Time Window* starting at time t as a two-dimensional array W of size $n \times n$ such that $W[i,j]$ represents the sum of the values generated by a source s_i within the $j-$th unitary time interval of W. That is $W[i,j] = \sum_{x:id_s(x)=i \wedge ts(x) \in \Delta t_j} value(x)$, where Δt_j is the time interval $[t+(j-1)\cdot u..t+j\cdot u]$. The whole data stream consists of an infinite sequence $W_1, W_2, \ldots$ of time windows such that the $i-$th one starts at $t_i = (i-1)\cdot T$ and ends at $t_{i+1} = i \cdot T$.

In the following, for the sake of presentation, we assume that the number of sources is a power of 2 (i.e. $n = 2^k$, where $k > 1$).

A *Quad-Tree Window* on the time window W, called $QTW(W)$, is a full 4−ary tree whose nodes are pairs $\langle r, sum(r)\rangle$ (where r is a block of W) such that:

1. $Root(QTW(W)) = \langle [1..n, 1..n], sum([1..n, 1..n])\rangle$;
2. each non leaf node $q = \langle r, sum(r)\rangle$ of $QTW(W)$ has four children representing the four quadrants of r; that is, $Child(q,i) = \langle r_i, sum(r_i)\rangle$ for $i = 1, \ldots, 4$.
3. the depth of $QTW(W)$ is $log_2 n + 1$.

Property 3 implies that each leaf node of $QTW(W)$ corresponds to a single element of the time window array W. Given a node $q = \langle r, sum(r)\rangle$ of $QTW(W)$, r is referred to as $q.range$ and $sum(r)$ as $q.sum$.

The space needed for storing all the nodes of a quad-tree window $QTW(W)$ is larger than the one needed for a flat representation of W. In fact, it can be easily shown that the number of nodes of $QTW(W)$ is $\frac{4\cdot n^2-1}{3}$, whereas the number of elements in W is n^2. Indeed, $QTW(W)$ can be represented compactly, exploiting the hierarchical structure of the quad-tree partition and the possible sparsity of data in a time window (i.e. the possible presence of null blocks in the quad-tree window). In [1] it has been shown that, if we use 32 bits for representing a sum, the largest storage space needed for a quad-tree window is $S^{max}_{QTW} = (32 + 8/3)n^2 - 2/3$ bits.

3.3 Populating Quad-Tree Windows

In this section we describe how a quad-tree window is populated as new data arrive. Let W_k be the time window associated to a given time interval $[(k-1)\cdot T..k\cdot T]$, and $QTW(W_k)$ the corresponding quad-tree window. Let $x = \langle id_s, v, ts\rangle$ be a new sensor reading such that ts is in $[(k-1)\cdot T..k\cdot T]$. We next describe how $QTW(W_k)$ is updated on the fly, to represent the change of the content of W_k.

Let $QTW(W_k)_{old}$ be the quad-tree window representing the content of W_k before the arrival of x. If x is the first received reading whose timestamp belongs to the time interval of W_k, $QTW(W_k)_{old}$ consists of a unique null node (the root). An algorithm for updating a quad-tree window on a reading arrival can work as follows. The algorithm takes as arguments x and $QTW(W_k)_{old}$, and returns the up-to-date quad-tree window Q_{new} on W_k. First, the old quad-tree window $QTW(W_k)_{old}$ is assigned to Q_{new}. Then, the algorithm determines the coordinates $\langle id_s, j\rangle$ of the element of W_k which must be updated according to the arrival of x, and visits Q_{new} starting from its root. At each step of the visit, the algorithm processes a node of Q_{new} corresponding to a block of W_k which contains $\langle id_s, j\rangle$The sum associated with the node is updated by adding $value(x)$ to it (see Fig. 4). If the visited node was null (before the updating), it is split into four new null children. After updating the current node (and possibly splitting it), the visit goes on processing the child of the current node which contains $\langle id_s, j\rangle$. The algorithm ends after updating the node of Q_{new} corresponding to the single element $\langle id_s, j\rangle$. The details of this algorithm (as well as all the other algorithms sketched in this paper) are reported in [1].

4. THE MULTI-RESOLUTION DATA STREAM SUMMARY

A quad-tree window represents the readings generated within a time interval of size T. The whole sensor data stream can be represented by a sequence of quad-tree windows $QTW(W_1), QTW(W_2), \ldots$ When a new sensor reading x arrives, it is inserted in the corresponding quad-tree window $QTW(W_k)$, where $ts(x) \in [(k-1)\cdot T..k\cdot T]$. A quad-tree window $QTW(W_k)$ is physically created when the first reading belonging to $[(k-1)\cdot T..k\cdot T]$ arrives.

In this section we define a structure that both indexes the quad-tree windows and summarizes the values carried by the stream. This structure is called *Multi-Resolution Data Stream Summary* and pursues two aims: 1) making range queries involving more than one time window efficient to evaluate; 2) making the stored data easy to compress.

We propose the following scheme for indexing quad-tree windows:

1. time windows are clustered into groups $C_1, C_2, \ldots$; each cluster consists of K contiguous time windows, thus describing a time interval of size $K \cdot T$;
2. quad-tree windows inside each cluster C_l are indexed by means of a binary tree denoted by $BTI(C_l)$;
3. the whole index consists of a list linking $BTI(C_1), BTI(C_2), \ldots$

We next focus our attention on describing the structure of a single index $BTI(C_l)$. Then, we show how the whole index overlying the quad-tree windows is built.

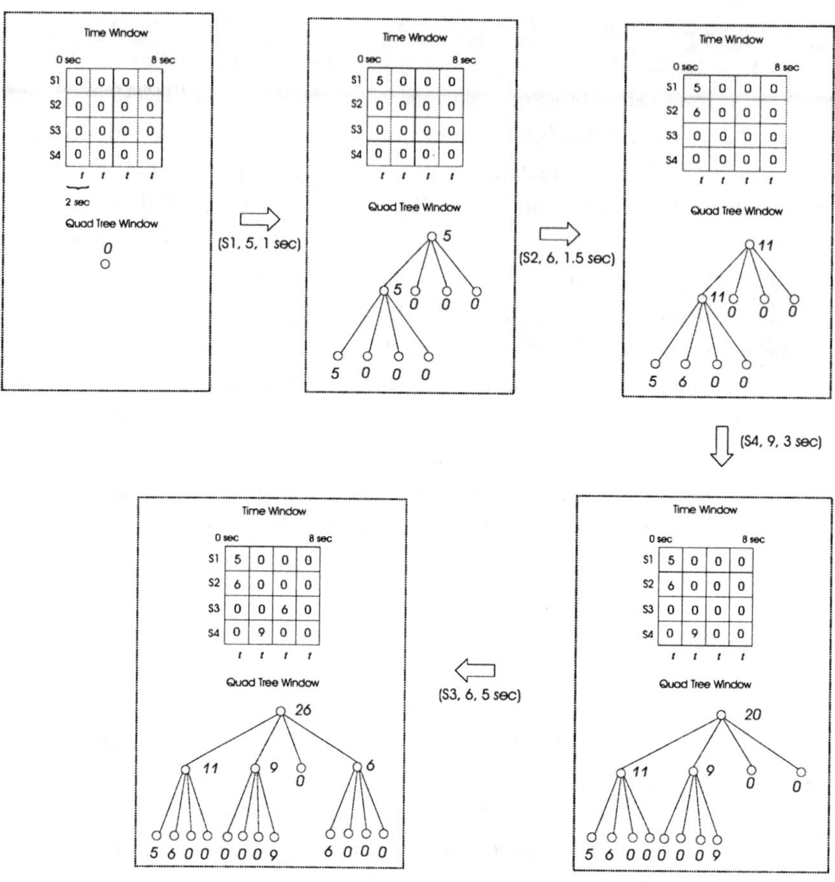

Fig. 4. Populating a quad-tree window.

4.1 Indexing a Cluster of Quad-Tree Windows

Consider the l-th cluster C_l of the sequence representing the whole sensor data stream. C_l corresponds to the time interval $[(l-1) \cdot K \cdot T..l \cdot K \cdot T]$. The time interval corresponding to C_l will be denoted by $\Delta T(C_l)$. We fix the value of K to a power of 2.

A *Binary Tree Index* on C_l, is denoted by $BTI(C_l)$ and is a full binary tree whose nodes are pairs $\langle t, s\rangle$, with t a time interval and s a sum, such that:

1. $Root(BTI(C_l)) = \langle \Delta T(C_l), sum(\Delta T(C_l))\rangle$ where $sum(\Delta T(C_l))$ is the sum of the values generated within $\Delta T(C_l)$ by all the sources, that is $sum(\Delta T(C_l)) = \sum_{(l-1)\cdot K<i\leq l\cdot K} sum(W_i)$

2. each non leaf node $q = \langle t, s \rangle$ of $BTI(C_l)$, with $t = [j_1T..j_2T]$, has two child nodes corresponding to the two halves of t, that is $Child(q,i) = \langle t_{i/2}, s_{i/2} \rangle, i = 1, 2$, where $t_{i/2}$ is the $i-$th half of t, and $s_{i/2}$ is the sum of all the readings generated within $t_{i/2}$ by all the sources.
3. the depth of $BTI(C_l)$ is log_2K, that is each leaf node of $BTI(C_l)$ corresponds to a time interval of size $2T$.
4. each leaf node $q = \langle t, s \rangle$ of $BTI(C_l)$, with $t = [j_1T..j_2T]$ $(j_2 - j_1 = 2)$, refers to the two quad-tree windows in t (i.e. $QTW(W_i), j_1 < i \leq j_2$).

Given a node $q = \langle t, s \rangle$ of $BTI(C_l)$, t and s are referred to as $q.interval$ and $q.sum$, respectively. Moreover $q.range$ denotes the range $\langle s_1..s_n, t \rangle$.
In the same way as quad-tree windows, binary tree indices can be stored in a compact fashion as well. The largest space consumption of a binary tree index (embedding its referred QTWs) can be shown to be $S_{BTI}^{max} = (32 + 8/3) \cdot K \cdot n^2 + (52/3) \cdot K - 2$ bits.

4.2 Constructing and Linking Binary Tree Indices

In the same way as quad-tree windows, binary tree indices can be constructed dynamically, as new data arrive and new quad-tree windows are created. An algorithm for constructing a binary tree index follows the same strategy as the algorithm described in Section 3.3, and, in particular, uses that algorithm for populating the indexed quad-tree windows. The resulting algorithm consists in a function which takes as arguments a "new" reading x and the binary tree index $BTI(C_l)$ where x is in $\Delta T(C_l)$, and updates both the index and the underlying quad-tree windows.

The overall index on the sensor data stream is obtained by linking together $BTI(C_1)$, $BTI(C_2)$, . . ., i.e. the binary tree indices corresponding to consecutive clusters. In particular, when a new sensor reading x arrives, it is inserted (according to the just described algorithm) into the binary tree index $BTI(C_l)$ such that $ts(x)$ is in $\Delta T(C_l)$. If this BTI does not exist (i.e. x is the first arrival in this cluster), first of all a new binary tree index $BTI(C_l)$ containing a unique null node (the root) is created. Then the function for inserting x into $BTI(C_l)$ is called and the updated BTI returned by that function is added to the existing list of consecutive binary tree indices. The list of BTIs with the underlying list of quad-tree windows is referred to as *Multi-Resolution Data Stream Summary* - MRDS. As the sensor data stream is infinite, the length of the list of binary tree indices is not bounded, so that a MRDS cannot be physically stored. In the following section we propose a compression technique which allows us to store the most relevant information carried by the (infinite) sensor data stream by keeping a finite list of (compressed) binary tree indices.

5. COMPRESSION OF THE MULTI-RESOLUTION DATA STREAM SUMMARY

Due to the bounded storage space which is available to store the information carried by the sensor data stream, the Multi-Resolution Data Stream Summary cannot be physically represented, as the stream is potentially infinite.

As new sensor readings arrive, the available storage space decreases till no other reading can be stored. Indeed, we can assume that recent information is more relevant than the older one for answering user queries, which usually investigate the recent evolution of the monitored world. Therefore, older information can be reasonably represented with less detail than recent data. This suggests us the following approach: as new readings arrive, if there is not enough storage space to represent them, the needed storage space is obtained by discarding some detailed information about "old" data.

We next describe our approach in detail. Let x be the new sensor reading to be inserted, and let $BTI(C_1), BTI(C_2), \ldots, BTI(C_k)$ be the list of binary tree indices representing all the sensor readings preceding x. This means that x must be inserted into $BTI(C_k)$. The insertion of x is done by performing the following steps:

1. the storage space $Space(x)$ needed to represent x into $BTI(C_k)$ is computed by evaluating how the insertion of x modifies the structure and the content of $BTI(C_k)$. $Space(x)$ can be easily computed using the same visiting strategy as the algorithm for inserting x into $BTI(C_k)$;
2. if $Space(x)$ is larger than the left amount $Space_a$ of available storage space, then the storage space $Space(x) - Space_a$ is obtained by compressing (using a lossy technique) the oldest binary tree indices, starting from $BTI(C_1)$ towards $BTI(C_k)$, till enough space is released.
3. x is inserted into $BTI(C_k)$.

We next describe in detail how the needed storage space is released from the list $BTI(C_1), BTI(C_2), \ldots, BTI(C_k)$. First, the oldest binary tree index is compressed (using a technique that will be described later) trying to release the needed storage space. If the released amount of storage space is not enough, then the oldest binary tree index is removed from the list, and the same compression step is executed on the new list $BTI(C_2), BTI(C_3), \ldots, BTI(C_k)$. The compression process ends when enough storage space has been released from the list of binary tree indices.

The strategy followed for compressing a single BTI (i.e. the oldest one of the list) exploits the hierarchical structure of the binary tree indices: each internal node of a BTI contains the sum of its child nodes, and the leaf nodes contain the sum of all the reading values contained in the referred quad-tree windows. This means that the information stored in a node of a BTI is replicated with a coarser "resolution" in its ancestor nodes. Therefore, if we delete two sibling nodes from a binary tree index, we do not lose all the information carried by these nodes: the sum of their values is kept in their ancestor nodes. Analogously, if we delete a quad-tree window QTW_k, we do not lose every information about the values of the readings belonging to the time interval $[(k-1) \cdot T..k \cdot T]$, as their sum is kept in a leaf node of the BTI.

As it will be described later, the compression of the oldest BTI is obtained by either compressing the referred QTWs (using an ad hoc technique for compressing quad-trees) or pruning some of the BTI nodes. This means that the compression process modifies the structure of a BTI:

- a Compressed BTI is not, in general, a full binary tree, as it is obtained from a full tree (i.e. the original BTI) by deleting some of its nodes;
- not every leaf node refers to two QTWs, as a leaf node of the compressed BTI can be obtained in three ways: 1) it corresponds to a leaf node of the original BTI; 2) it corresponds to a leaf node of the original BTI whose referred QTWs have been deleted; 3) it corresponds to an internal node of the original BTI whose child nodes have been deleted.

We next describe in detail the compression process of a BTI. The BTI to be compressed is visited in order to reach the left-most node N (i.e. the oldest node) having one of the following properties:

1. N is a leaf node of the BTI which refers to 2 QTWs;
2. the node N has 2 child leaf nodes, and all the 2 children do not refer to any QTW.

In the first of the two cases, an ad hoc procedure for compressing the quad-tree windows referred by N is called. The 2 QTWs are compressed till either the needed storage space is released, or they cannot be further compressed. If both QTWs are no longer compressible, then they are deleted definitively. In the second of the two cases, the children of N are deleted. The information contained in these nodes is kept summarized in N.
In Fig. 5, several steps of the compression process on a binary tree index of depth 4 (i.e. a BTI indexing 16 QTWs) are shown.

The QTWs underlying the BTI are represented by squares. In particular, uncompressed QTWs are white, partially compressed are grey, whereas QTWs which cannot be further compressed are crossed. We next describe the compression process reported in Fig. 5. At step 1, the oldest QTW is partially com-

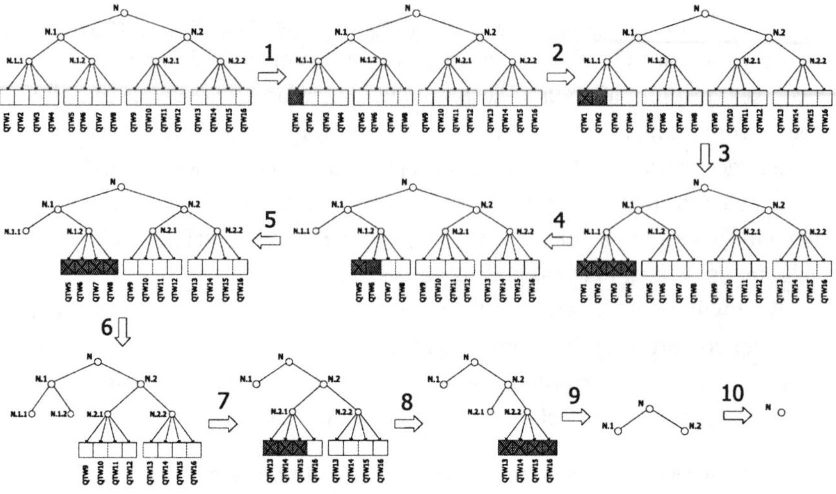

Fig. 5. Compressing a BTI

pressed. At step 2, the needed storage space is released by continuing the compression of QTW_1 till it cannot be further compressed. As the released storage space is not enough, QTW_2 is partially compressed. After step 3, all the QTWs referred by $N.1.1$ are maximally compressed, and they are removed during step 4. Step 6 consists of removing the four QTWs referred by $N.1.2$. The node $N.1.2$ will be removed together with $N.1.1$ during step 7: as the space released by deleting $N.1.1$ and $N.1.2$ does not suffice, some QTWs referred by $N.2.1$ are compressed too during the same compression step. The compression process ends after step 10: the BTI consists of a unique node (the root) which will be definitively removed as further storage space is needed.

The compression of a BTI consists in removing its nodes progressively, so that the detailed information carried by the removed nodes is kept summarized in their ancestors. This summarized data will be exploited (as described in Section 6) to estimate the original information represented in the removed QTWs underlying the BTI. The depth of a BTI (or, equivalently, the number of QTWs in the corresponding cluster) determines the maximum degree of aggregation which is reached in the MRDS. This parameter depends on the application context. That is, the particular dynamics of the monitored world determines the average size of the time intervals which need to be investigated in order to retrieve useful information. Data summarizing time intervals which are too large w.r.t. this average size are ineffective to exploit in order to estimate relevant information. For instance, the root of a BTI whose depth is 100 contains the sum of the readings produced within 2^{100} consecutive time windows. Therefore, the value associated to the root cannot be profitably used to estimate the sum of the readings in a single time window effectively (unless additional information about the particular data distribution carried by the stream is available).

5.1 Compressing Quad-Tree Windows

The strategy used for compressing binary tree indices could be adapted for compressing quad-tree windows. For instance, we could compress a quad-tree window incrementally (i.e. as new data arrive) by searching for the left-most node N having 4 child leaf nodes, and then deleting these children.

Indeed, we refine this compression strategy in order to delay the loss of detailed information inside a QTW. Instead of simply deleting a group of nodes, we try to release the needed storage space by replacing their representation with a less accurate one, obtained by using a lower numeric resolution for storing the values of the sums. To this end, we use a compact structure (called *n Level Tree index - nLT*) for representing approximately a portion of the QTW. nLT indices were first proposed in [5,6], where they are shown to be very effective for the compression of two-dimensional data. A nLT index occupies 64 bits and describes approximately both the structure and the content of a sub-tree with depth at most n of the QTW. An example of nLT index (called "*3 Level Tree index*" - 3LT) is shown in Fig. 6. The left-most sub-tree $SQTW$ of the quad-tree of this figure consists of 21 nodes, which occupy $2 \cdot 21 + 32 \cdot 16 = 554$ bits ($2 \cdot 21$ bits are used to represent their structure, whereas $32 \cdot 16$ bits to represent the sums of all non derivable nodes). The 64 bits of the nLT index used for $SQTW$ are organized as follows: the first 17 bits are used to represent the second level of $SQTW$, the second 44 bits for the third level, and the remainder 3 bits for some structural information about the index. That is, the four nodes in the second level of $SQTW$ occupy $3 \cdot 32 + 4 \cdot 2 = 104$ bits in the exact representation, whereas they consume only 17 bits in the index. Analogously, the 16 nodes of the third level of $SQTW$ occupy $4 \cdot (3 \cdot 32 + 4 \cdot 2) = 416$ bits, and only 44 bits in the index. In Fig. 6 the first 17 bits of the 3LT index are described in more detail.

Two strings of 6 bits are used for storing $A.sum + B.sum$ and $A.sum + C.sum$, respectively, and further 5 bits are used to store $A.sum$. These string of bits do not represent the exact value of the corresponding sums, but they represent the sums as fractions of the sum of the parent node. For instance, if $R.sum$ is 100 and $A.sum = 25$, $B.sum = 30$, the 6 bit string representing $A.sum + B.sum$ stores the value: $L_{A+B} = round\left(\frac{A.sum+B.sum}{R.sum} \cdot (2^6 - 1)\right) = 35$,

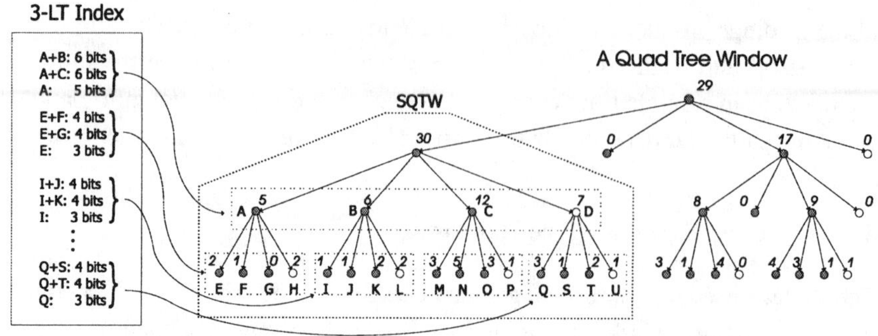

Fig. 6. A 3LT index associated to a portion of a quad-tree window.

whereas the 5 bit string representing $A.sum$ stores the following value: $L_A = round\left(\frac{A.sum}{A.sum+B.sum} \cdot (2^5 - 1)\right) = 14$. An estimate of the sums of A, B, C, D can be evaluated from the stored string of bits. An estimate of $A.sum + B.sum$ is given by: $\overline{A.sum + B.sum} = \frac{L_{A+B}}{2^6-1} \cdot R.sum = 55.6$, whereas an estimate of $B.sum$ is computed by subtracting the estimate of $A.sum$ (obtained by using L_A) from the latter value. The 44 bits representing the third level of $SQTW$ are organized in a similar way. For instance, two strings of 4 bits are used to represent $E.sum + F.sum$ and $E.sum + G.sum$, respectively, and a string of 3 bits is used for $E.sum$. The other nodes at the third level are represented analogously. We point out that saving one bit for storing the sum of A w.r.t. $A + B$ can be justified by considering that, on average, the value of the sum of the elements inside A is half of the sum corresponding to $A + B$, since the size of A is half of the size of $A + B$. Thus, on the average, the accuracy of representing $A + B$ using 6 bits is the same as the accuracy of representing A using 5 bits.

The family of nLT indices includes several types of index other than the 3LT one. Each of these indices reflects a different quad-tree structure: 3LT describes a balanced quad-tree with 3 levels, 4LT (*4 Level Tree*) an unbalanced quad-tree with at most 4 levels, and so on. However, the exact description of nLT indices is beyond the aim of this paper. The detailed description of these indices can be found in [6]. The same portion of a quad-tree window could be represented approximately by any of the proposed nLT indices. In [6] a metric for choosing the most "suitable" nLT index to approximate a portion of a quad-tree is provided: that is, the index which permits us to re-construct the original data distribution most accurately. As it will be clear next, this metric is adopted in our compression technique: the oldest "portions" of the quad-tree window are not deleted, but they are replaced with the most suitable nLT index. The algorithm which uses indices to compress a QTW is analogous to the algorithm for compressing a BTI (suitably adapted to work with 4-ary trees) sketched in Section 5. That is the QTW to be compressed is visited in order to reach the left-most node N (i.e. the oldest node) having one of the following properties: 1) N is an internal node of the QTW such that $size(N.range) = 16$; 2) the node N has 4 child leaf nodes, and each child is either null or equipped with an index. Once the node with one of these properties is found, it is equipped with the most suitable nLT index, and all its descending nodes are deleted. In particular, in case 1 (i.e. N is two from the last level of the uncompressed QTW) N is equipped with a $3LT$ index. In case 2 the following steps are performed: 1) all the children of N which are equipped with an index are "expanded": that is, the quad-trees represented by the indices are approximately re-constructed; 2) the most suitable nLT index I for the quad-tree rooted in N is chosen, using the above cited metric [6]; and 3) N is equipped with I and all the nodes descending from N are deleted.

6. ESTIMATING RANGE QUERIES ON A MULTI-RESOLUTION DATA STREAM SUMMARY

A sum range query $Q = \langle s_i..s_j, [t_{start}..t_{end}]\rangle$ can be computed by summing the contributions of every QTW corresponding to a time window overlapping $[t_{start}..t_{end}]$. The QTWs underlying the list of BTIs are represented by means of a linked list in time ascending order. Therefore the sub-list of QTWs giving some contribution to the query result can be extracted by locating the first (i.e. the oldest) and the last (i.e. the most recent) QTW involved in the query (denoted, respectively, as QTW_{start} and QTW_{end}). This can be done efficiently by accessing the list of BTIs indexing the QTWs, and locating the first and the last BTI involved in the query. That is, the binary tree indices BTI_{start} and BTI_{end} which contain a reference to QTW_{start} and QTW_{end}, respectively. BTI_{start} and BTI_{end} can be located efficiently, by performing a binary search on the list of BTIs. Then, QTW_{start} and QTW_{end} are identified by visiting BTI_{start} and BTI_{end}. The answer to the query consists of the sum of the contributions of every QTW between QTW_{start} and QTW_{end}. The evaluation of each of these contributions is explained in detail in the next section.

Indeed, as the Sensor Data Stream Summary is progressively compressed, it can happen that QTW_{start} has been removed, and the information it contained is only represented in the overlying BTI with less detail. Therefore, the query can be evaluated as follows: 1) the contribution of all the removed QTWs is estimated by accessing the content of the nodes of the BTIs where these QTWs are summarized and 2) the contribution of the QTWs which have not been removed is evaluated after locating the oldest QTW involved in the query which is still stored. This QTW will be denoted as QTW'_{start}.

Indeed, it can happen that QTW_{end} has been removed either. This means that all the QTWs involved in the query have been removed by the compression process to release some space, as the QTWs are removed in time ascending order. In this case, the query is evaluated by estimating the contribution of each involved QTW by accessing only the nodes of the overlying BTIs.

For instance, consider the MRDS consisting of two BTIs shown in Fig. 7. The QTWs and the BTI nodes whose perimeter is dashed have been removed by the compression process. The query represented with a grey box is evaluated by summing the contributions of the BTI_1 node $N1.1$ with the contribution of each QTW belonging to the sequence $QTW_9, QTW_{10}, \ldots, QTW_{29}$. The

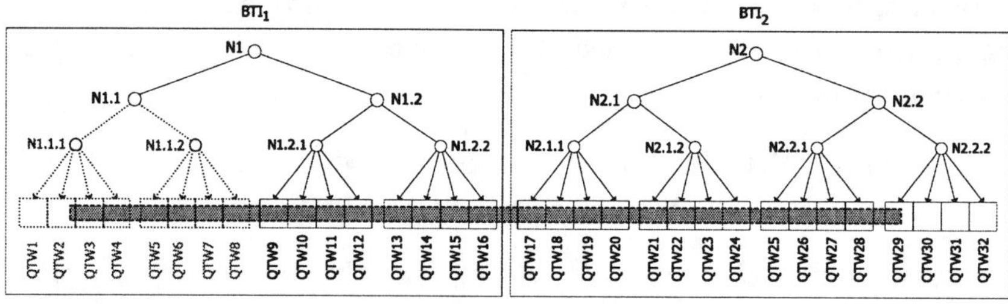

Fig. 7. A range query on a MRDS

query estimation algorithm uses a function $BTIBinarySearch$ which takes as arguments a Multi-Resolution Data Stream Summary and the time boundaries of the range query, and returns the first and the last BTI of the summary involved in the query. Moreover, it uses the function $EstimateAndLocate$. This function is first invoked on BTI_{start} and performs two tasks: 1) it evaluates the contribution of the BTI nodes involved in the query where the information of the removed QTWs is summarized, and 2) it locates (if possible) QTW'_{start}, i.e. the first QTW involved in the query which has not been removed. If QTW'_{start} is not referred by BTI_{start}, $EstimateAndLocate$ is iteratively invoked on the subsequent BTIs, till either QTW'_{start} is found or all the BTIs involved in the query have been visited. The contribution of the BTI leaf nodes to the query estimate is evaluated by performing linear interpolation. The use of linear interpolation on a leaf node N of a BTI is based on the assumption that data are uniformly distributed inside the two-dimensional range $N.range$ (CVA - *Continuous Value Assumption*). If we denote the two dimensional range corresponding to the intersection between $N.range$ and the range of the query Q as $N \cap Q$, and the size of the whole two dimensional range delimited by the node N as $size(N)$, the contribution of N to the query estimate is given by: $\frac{size(N \cap Q)}{size(N)} \cdot N.sum$.

6.1 Estimating a Sum Range Query inside a QTW

The contribution of a QTW to a query Q is evaluated as follows. The quad-tree underlying the QTW is visited starting from its root (which corresponds to the whole time window). When a node N is being visited, three cases may occur:

1. *the range corresponding to the node is external to the range of* Q: the node gives no contribution to the estimate;
2. *the range corresponding to the node is entirely contained into the range of* Q: the contribution of the node is given by the value of its sum;
3. *the range corresponding to the node partially overlaps the range of* Q: if N is a leaf and is not equipped with any index, linear interpolation is performed for evaluating which portion of the sum associated to the node lies onto the range of the query. If N has an index, the index is "expanded"(i.e. an approximate quad-tree rooted in N is re-constructed using the information contained in the index). Then the new quad-tree is visited with the same strategy as the QTW to evaluate the contribution of its nodes (see [6] for more details). Finally, if the node N is internal, the contribution of the node is the sum of the contributions of its children, which are recursively evaluated.

The pre-aggregations stored in the nodes of quad-tree windows make the estimation inside a QTW very efficient. In fact, if a QTW node whose range is completely contained in the query range is visited during the estimation process, its sum contributes to the query result exactly, so that none of its descending nodes must be visited. This means that not all the leaf nodes involved in the query need to be accessed when evaluating the query estimate. The overall estimation process turns out to be efficient thanks to the hierarchical organization of data in the QTWs, as well as the use of the overlying BTIs which permits us to locate the quad-tree windows efficiently. We point out that the BTIs involved in the query can be located efficiently too, i.e. by performing a binary search on the ordered list of BTIs stored in the MRDS. The cost of this operation is logarithmic with respect to the list length, which is, in turn, proportional to the number of readings represented in the MRDS.

6.2 Answering Continuous (Range) Queries

The range query evaluation paradigm on the data summary can be easily extended to deal with *continuous range queries*. A *continuous query* is a triplet $Q = \langle s_i..s_j, \Delta T_{start}, \Delta T_{end} \rangle$ (where $\Delta T_{start} > \Delta T_{end}$) whose answer, at the current time t, is the evaluation of an aggregate operator (such as *sum, count, avg*, etc.) on the values produced by the sources $s_i, s_{i+1}, \ldots, s_j$ within the time interval $[t - \Delta T_{start}..t - \Delta T_{end}]$. In other words, a continuous query can be viewed as a range query whose time interval "moves" continuously, as time goes on. The output of a continuous query is a stream of (simple) range query answers that are evaluated with a given frequency. That is, the answer to a continuous query $Q = \langle s_i..s_j, \Delta T_{start}, \Delta T_{end} \rangle$ issued at time t_0 with frequency Δt is the stream consisting of the answers of the queries $Q_0 = \langle s_i..s_j, t_0 - \Delta T_{start}, t_0 - \Delta T_{end} \rangle, Q_1 = \langle s_i..s_j, t_0 - \Delta T_{start} + \Delta t, t_0 - \Delta T_{end} + \Delta t \rangle, Q_2 = \langle s_i..s_j, t_0 - \Delta T_{start} + 2 \cdot \Delta t, t_0 - \Delta T_{end} + 2 \cdot \Delta t \rangle, \ldots$. The i-th term of this stream can be evaluated efficiently if we exploit the knowledge of the $(i-1)$-th value of the stream, provided that $\Delta t \ll \Delta T_{start} - \Delta T_{end}$. In this case the ranges of two consecutive queries Q_{i-1} and Q_i are overlapping, and Q_i can be evaluated by answering two range queries whose size is much less than the size of Q_i. These two range queries are $Q' = \langle s_i..s_j, t_0 - \Delta T_{start} + (i-1) \cdot \Delta t, t_0 - \Delta T_{start} + i \cdot \Delta t \rangle$, and $Q'' = \langle s_i..s_j, t_0 - \Delta T_{end} + (i-1) \cdot \Delta t, t_0 - \Delta T_{end} + i \cdot \Delta t \rangle$. Thus we have: $Q_i = Q_{i-1} - Q' + Q''$.

REFERENCES

1. E. Masciari, D. Sacca, A. Cuzzocrea, F. Furfaro and C. Sirangelo. Approximate Query Answering on Sensor Network Data Streams. Technical Report 9, ICAR-CNR, 2003. Available at `http://www.icar.cnr.it/isi/`.
2. M. Datar, R. Motwani, J. Widom, B. Babcock and S. Babu. Models and Issues in Data Stream Systems. In *Proc. PODS Symp.*, 2002.
3. D. Pregibon, A. Rogers, F. Smith, C. Cortes and K. Fisher. Hancock: A Language for Extracting Signatures from Data Streams. In *Proc. KDD Conf.*, 2000.
4. V. Tsotras, B. Seeger, D. Zhang and D. Gunopulos. Temporal Aggregation over Data Streams Using Multiple Granularities. In *Proc. EDBT Conf.*, 2002.
5. D. Rosaci, D. Sacca, F. Buccafurri and L. Pontieri. Improving Range Query Estimation on Histograms. In *Proc. ICDE*, 2002.
6. D. Sacca, C. Sirangelo, F. Buccafurri and F. Furfaro. A Quad-Tree Based Multiresolution Approach for Two-Dimensional Summary Data. In *Proc. SSDBM Conf.*, 2003.
7. S. Rajagopalan, M. Henzinger and P. Raghavan. Computing on Data Streams. Technical Report 1998-011, Digital Systems Research Center, 1998. Available at `http://www.research.digital.com/SRC/`.
8. J. M. Hellerstein and R. Avnur. Eddies: Continuously Adaptive Query Processing. In *Proc. ACM SIGMOD Conf.*, 2000.
9. M. J. Franklin and S. Madden. Fjording the Stream: An Architecture for Queries over Streaming Sensor Data. In *Proc. ICDE*, 2002.

Georouting and Delta-Gathering: Efficient Data Propagation Techniques for GeoSensor Networks

Dina Goldin, Mingjun Song, Ayferi Kutlu, Huayan Gao, and Hardik Dave

Dept. of Computer Science and Engineering
University of Connecticut
Storrs, CT 06269, USA

ABSTRACT

We consider the issue of *query and data propagation* in the context of geosensor networks over geo-aware sensors. In such networks, techniques for efficient propagation of queries and data play a significant role in reducing energy consumption.

Georouting is a new technique for the broadcasting of *localized data and queries* in geo-aware sensor networks; it makes use of the existing query routing tree, and does not involve the creation of any additional communication channels. In addition to localized broadcasting, georouting is useful for (non-localized) broadcasting *spatial data*, greatly reducing the amount of communication, and hence energy consumption, during broadcasts. We demonstrate its effectiveness empirically, having implemented this technique.

In addition to broadcasting queries and data to the sensors, we consider *data gathering*, where data is being transmitted from the sensors back towards the central processor. *Delta-gathering* is a new technique for reducing the amount of communication during data gathering.

Finally, we apply our delta-gathering approach toward the problem of sensor data *visualization*. We present *sensor terrains* as a preferable alternative to isoline-based visualization (*contour maps*) for this problem.

1 INTRODUCTION

Sensor networks can be embedded in a variety of geographic environments, such as high-rise buildings, airports, highway stretches, or even the ocean. They enable the monitoring of these environments for a wide variety of applications, from security to biological. For many of the anticipated applications, the ability to query sensor networks in an *ad hoc* fashion is key to their usefulness. Rather than re-engineering the network for every task, as is commonly done now, *ad hoc querying* allows the same network to process any of a broad class of queries, by expressing these queries in some query language. In essence, the network appears to the user as a single distributed agent whose job it is to observe the environment wherein it is embedded, and to interact with the user about its observations.

Unlike traditional database applications, where spatial considerations are often irrelevant (except as expressed by traditional attributes such as *address* or *zip code*), it is believed that most applications of sensor networks, in such diverse fields as security, civil engineering, environmental engineering, or meteorology, will involve queries that combine *spatial data* with *streaming sensor data*. For this reason, we are focusing our investigation on a query system that combines a *spatial database* [26] with a *geo-aware sensor network* [11] SPASEN-QS for short. There are currently several research projects, including those at Berkeley [22, 23, 25] and Cornell [34, 35] dealing with query issues in sensor networks. However, we are not aware of any other projects that have focused on sensor network querying for spatial data.

As is common for the sensor network query setting, SPASEN-QS architecture involves a *central processor* which hosts the spatial data and provides a user interface to the query system. A *routing tree* is maintained over the sensors, whose root communicates directly with the central processor. All communication is therefore *vertical*, either down from the central processor towards the sensors (*broadcasting*, or *distributing*) or up from the individual sensors towards the central processor (*gathering*, or *collecting*).

Sensors are expected to run battery-powered and unattended for long periods of time, hence the need to minimize their *energy consumption*. Energy consumption therefore serves as the *optimization metric* for sensor network computations, analogous to *time and space complexity* in traditional computation.

Of the four types of sensor activities (*transmitting*, *sensing*, *receiving*, *computing*), the first is the most *expensive* in terms of energy consumption. Efficient techniques for the propagation of queries and data in sensor networks play a significant role in reducing energy consumption for sensor network computation.

In this paper, we consider the issue of query and data propagation in geosensor network query systems such as SPASEN-QS. *Georouting* and *Delta-gathering* are the two techniques we propose.

Georouting is a new technique for localized broadcasting of queries in geo-aware sensor networks; it makes use of the existing query routing tree, and does not involve the creation of any additional communication channels. Besides localized query broadcasting, georouting is also useful when broadcasting spatial data, greatly reducing the amount of communication, and hence energy consumption, during broadcasts. We have implemented georouting, and demonstrate its effectiveness empirically.

In addition to broadcasting queries and data to the sensors, we consider *data gathering*, where data movement is reversed towards the central data manager.

Delta-gathering is an new technique for reducing the amount of communication during data gathering. The goal of delta-gathering is to improve power consumption of the sensor network by reducing the amount of communication at the gathering phase. In the absence of a new value from some sensor, unless we know that the sensor is down, we assume that the value at this sensor has not appreciably changed since the last transmission, and is not worth transmitting. Note that this technique does not affect the *semantics* of the data, only the method of gathering.

We apply delta-gathering toward the problem of *sensor data visualization* via *sensor terrains*. Sensor terrains are a preferable alternative to isoline-based visualization [12]. They are represented by *triangulated irregular networks* (TINs). Visualization of sensor terrains is therefore a special case of *dynamic TIN generation*, a computational geometry problem for which we present a new incremental delta-based algorithm.

At any given time t, each sensor in the network corresponds to a point (x, y, z), where (x, y) is the location of the sensor and z is its reading at time t. A *sensor terrain* is a surface which passes through all these sensor points. As the readings change, so does the sensor terrain; it is *dynamic*, more like a video than a static surface. There are several reasons to prefer sensor terrains to contours as the means of sensor data visualization: *more intuitive*, *less lossy*, *greater manipulability*, *easier updates*. These are discussed in section 3.

We represent sensor terrains by *triangulated irregular networks* (TINs) [7]; An alternative representation are NURBS [27]. For sensor data visualization, we must continuously regenerate the TIN corresponding to the dynamic sensor terrain. *Efficient dynamic TIN generation* is a new computational geometry problem for which we present an *incremental* $O(log\ n)$ algorithm.

Given a sensor terrain, a contour map can be computed from it (but not vice versa). We therefore conclude by presenting a new efficient algorithm for dynamically generating isolines from the sensor terrain.

Outline. We discuss georouting in section 2, sensor terrains in section 3, and isoline extraction in section 4. We conclude in section 5.

2 GEOROUTING

In this section, we discuss *georouting*, a new technique for localized broadcasting of queries in geo-aware sensor networks. In addition to localized query broadcasting, georouting is also useful when broadcasting spatial data, greatly reducing the amount of communication during broadcasts. We demonstrate its effectiveness empirically, and show that the use of special trees customized for georouting do not offer significant advantages over the existing routing tree.

2.1 Localized Broadcasting

In geospatial sensor networks, the data or the queries to be broadcast are often *localized*, i.e. of relevance only to those sensors located within a specific geographic region. When the information to be broadcast is spatial, the geolocation of the sensor often determines whether this information is relevant to it. For example, if a query needs to initialize sensors that are located within a given region X, then this operation is not relevant to those sensors which fall outside X; moreover, if all the sensors in a given subtree of the routing tree are outside of X, the information about X need not be routed to that subtree at all. Since communication consumes a large fraction of a sensor network's energy [33, 4], it is desirable to avoid unnecessary routing of spatial information.

Previous work on constraining the broadcasts to a geographic area include work in *geoaware routing* [14, 36], *directed diffusion* [13], rumor routing [1]. These algorithms were developed outside the sensor network querying context; they do not use a routing tree, relying on localized neighbor selection to efficiently route a packet to a destination. In contrast to these approaches, georouting relies on the existing *routing tree* for *all* communication. Specifically, it tags each node of the routing trees with *bounding box* information for itself and all its children. Furthermore, neither directed diffusion nor rumor routing make any use of geoinformation. Whereas the gradient information allows the localization of the sink node, messages in the opposite direction (from the sink) cannot be localized and involve a broadcast to all the nodes. *SRT trees* [22] have also been used for localized broadcasting, and are the most alike georouting trees. Both SRT and georouting trees involve decorating the existing query routing tree with additional information, without creating any additional communication channels. However, SRT trees store exactly one interval per attribute per node, whereas georouting trees store the intervals of each child as well. This results in much greater communication efficiency during localized broadcasts.

In addition, georouting is completely *decentralized*; the route is computed in-network rather than at the central processor. This is accomplished by augmenting the routing tree to make it geo-aware: at each internal node, the spatial bounding box of each child is stored; this bounding box is used during the routing to minimize unnecessary communication. We discuss the details of this algorithm in the next section.

2.2 Georouting Tree

Routing trees are more attractive for sensor network querying that in the standard network setting, due to the following three points of contrast between these settings:

- Normally, the sensor nodes serve strictly to route messages, with no in-network processing. In SNQ, there is in-network processing performed at the sensors to optimize query evaluation. Hence, SNQ nodes need to choose a single parent when routing data towards the sink, rather than send the same message to multiple candidate parents.

- Normally, the *sink* node, towards which the message is routed, changes often and a single tree routed at the sink cannot be maintained for long. In SNQ, a fixed root is assumed, which serves as the sink throughout the continuous evaluation of the query.

- While conversations in regular sensor networks between a source and a sink are short-lived (just long enough to send all the packets), sensor network queries are long-lived. They can perform monitoring functions over days if not months, during which time we must collect data continuously over the same path.

For the above reasons, a single routing tree that can be maintained over time, is the most suitable approach to routing in the case of SNQ.

Georouting trees augment routing tree architecture by maintaining at each sensor X a *bounding box* for each child Y of X, where a bounding box for Y encloses the geo-locations of all the sensors in the routing subtree rooted at Y. The bounding box of X is defined recursively as the maximum bounding rectangle of the bounding boxes for all of X's children, and the bounding box for each leaf node is simply its geo-location coordinates.

The algorithm for building the georouting tree is described next, based on original routing tree algorithms in [22, 23].

Algorithm for building the georouting tree:

1. (Assign levels top-down.) We assign a level to each node according to its distance from the root, starting by assigning 0 to the root itself. Given a current node A at level k in the tree, any node B within A's sensing range is assigned level $k + 1$ and added to the list of A's *candidate children*, unless it has already been assigned level k or less. Note that a node may be the candidate child of several nodes, each of which will be its *candidate parent*.
2. (Select the parents and compute the bounding boxes bottom-up.) Starting from the leaf nodes, we select one parent for each node, out of its list of candidate parents. We always select the geographically nearest node as the parent. Once a node's parent is chosen, we remove this node from the candidate children list of all other candidate parents.
3. (Assign the bounding box.) This operation is also done recursively, at the same time as step 2 (parent selection). First, assign the bounding box of

all leaf nodes to be their coordinate points and then goup to the root, calculate the bounding box of each node as the minimum rectangle which includes the bounding boxes of all its children. Store the bounding boxes of the children in the parents.

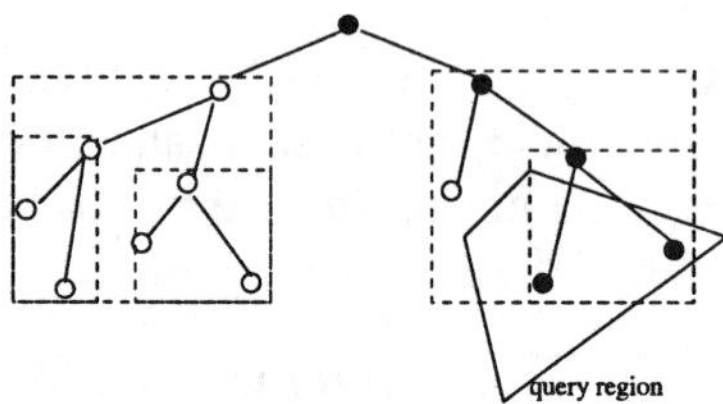

Figure 1: Message broadcast in georouting tree.

After building the georouting tree, the bounding box information at each internal node is used to *filter out* queries; the query is only transmitted to those children whose bounding boxes overlap with it. This is illustrated in figure 1. In this figure, the query region is on the right, and the bounding boxes are shown in dashed lines; the sensors where the query was routed are filled in, while the ones where the query was filtered out are white.

2.3 Georouting Tree Maintenance

Although in our setting we assume that the sensor nodes are not mobile, we cannot assume that the routing tree will stay constant over the duration of a query. This is due to the inherently dynamic nature of sensor networks, involving node failures, new nodes joining the network, etc. In this section, we analyze the communication cost of georouting tree updates. We do *not* consider here the costs incurred by the maintenance of the routing tree itself, but only on the additional costs needed to properly maintain the the bounding box information associated with the georouting tree.

Whenever a node joins or leaves the network, the geourouting tree needs to be updated; the update operations are *insert* and *delete*, respectively. For each operation, the bounding box of the node's parent needs to be recomputed. If the parameters of the parent's bounding box are changed, the parent's parent also has to be recomputed, and so on. Furthermore, if a non-leaf node fails, its children have to find new parents whose bounding boxes must be recomputed in a similar fashion.

In the best case, when a leaf node fails and its parents' bounding box is not affected, no messages may be needed to "repair" the tree. As soon as the parent node detects that it has not heard from its child for a period of time, it will remove that child's bounding box from its own without any messages involved. This is due to the fact that a geourouting tree node stores all of its children's bounding boxes locally. (For more information on how parents may detect the loss of a child, we refer to [23].) However, for an insert operation, there is at least one message involved, since the location of the new node must be communicated to its parent.

Let the parameter k represent the communication cost, for a random node s, of repairing all its ancestors in case of s's failure; $0 \leq s \leq d$, the depth of the tree. If the node to be deleted has children, the total communication costs are greater than k: the failure not only affects s's ancestors, but also the future ancestors of its children, who now need to select new parents. Each child needs at least one message to transmit its location to its new parent, plus k possible messages to propagate that change. The cost for each child is therefore the same as in case of *insert*, i.e. $k + 1$. The total cost for a deletion is therefore $k + c(k + 1)$, where c is the number of children of a failed node; the total cost for an insertion is $k + 1$.

To evaluate the communication cost of georouting tree updates, we performed an experiment to measure the following:

> When a random sensor node s is removed from the georouting tree, what is the average number of messages needed to update the tree?

This corresponds to $k + c(k + 1)$ in the above analysis.

Our experimental setting consisted of 1000 sensors with randomly assigned locations in a 100×100 area; the sensing range varied from 10 to 50, in steps of 5. After creating a georouting tree with a given sensing range, we simulated failure of a randomly chosen node by removing it from the tree, and performed a tree update, counting the number of messages. This number was averaged over many trials, to obtain the average total cost of deletion in a georouting tree.

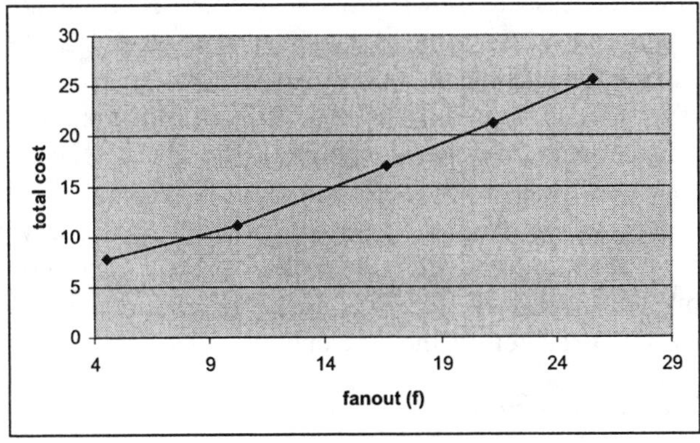

Figure 2: The cost of deletion in a georouting tree.

Figure 2 plots this cost against the *fanout* of the tree, i.e. the average number of children per internal node. We achieved higher fanouts by increasing the range while keeping the number of sensors fixed.

We conclude this section by noting that, in order to obtain communication savings from a georouting tree, it must be the case that tree updates do not occur too frequently. Specifically, if the expected cost of an update is c_u and the expected savings per epoch are c_s, then updates should occur on the average less than once per c_u/c_s epochs. We expect that this will be the case for many applications.

2.4 Experimental Results for Georouting

Having analyzed the costs associated with maintaining the georouting tree, we now consider the communication savings associated with georouting. In this section, we discuss an experiment that we have performed to access the performance of georouting, when compared either with SRT trees or with regular broadcasting. We report very significant savings, when compared with either of the other methods.

After choosing a fixed range of $(0, 100)$ in both x and y directions as the coordinate space of our "world" we randomly generated 1000 pairs of values in this range to simulate the positions of sensors. We then constructed a georouting tree over these sensors, with the root in the center of the world. Figure 3, generated automatically by our simulation, shows the georouting tree we obtained; here, the sensing range is set at 10 units.

We then simulated 500 localized broadcasts over this sensor network. For each broadcast, a rectangle was used to approximate the spatial region of interest (*query box*); this query box was generated randomly and propagated down the georouting tree. Figure 3 shows one such query box on the left; the paths involved in this broadcast are shown with thicker lines. Note that not all of these paths lead into the query box; some of them lead to nodes outside the query box, whose bounding boxes overlap the query box.

For each broadcast, the number of *hops* was measured and plotted against the number of sensors in the query box; figure 4 shows the resulting plot.

Analysis. We define *georouting efficiency* as the ratio between the minimum number of necessary hops from the root to all sensors in the query box and the number of hops used in georouting. We calculated that over 500 queries, the average number of necessary hops was 192, whereas the average number of actual hops was 229. Therefore, the efficiency is:

192/229 * 100% = 84%.

We ran exactly the same set of experiments using an SRT tree instead of a georouting tree. That is, each node only stored its own bounding box and not

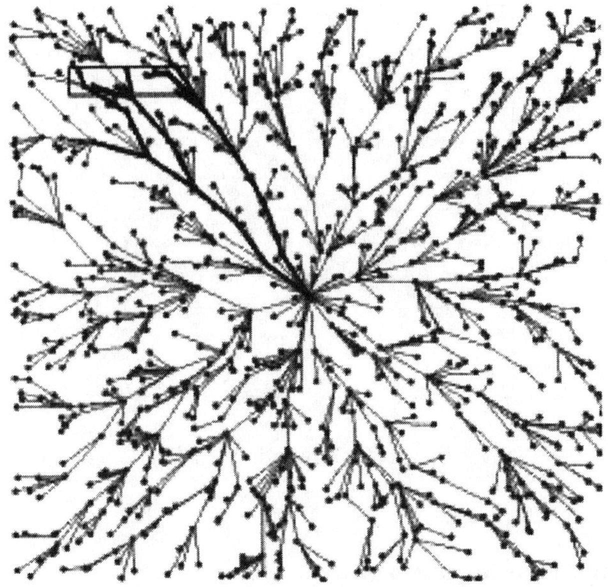

Figure 3: Georouting tree for our simulation.

the ones for its children. As a result, the average number of hops was 305, and the efficiency is much lower:

192/305 * 100% = 63%.

The above analysis measures how far georouting is from optimal routing. We can also compare georouting to regular tree routing, and measure what percentage of hops was saved. Regular tree routing would always result in 999 hops (one for every edge in the routing tree), whereas the average number of hops for our system was 229. Therefore, the percentage of hops saved is:

(999-229)/999 * 100% = 77%

Again, this is a significant improvement over the results for SRT routing:

(999-305)/999 * 100% = 69%

Furthermore, this saving can be compared with the cost of georouting tree updates in case of node failure or a new node joining the network. While that cost depends of the fanout (section 2.3), it is clear from our experiments that the savings with even a single broadcast of a localized query are greater than the cost of multiple updates to the tree.

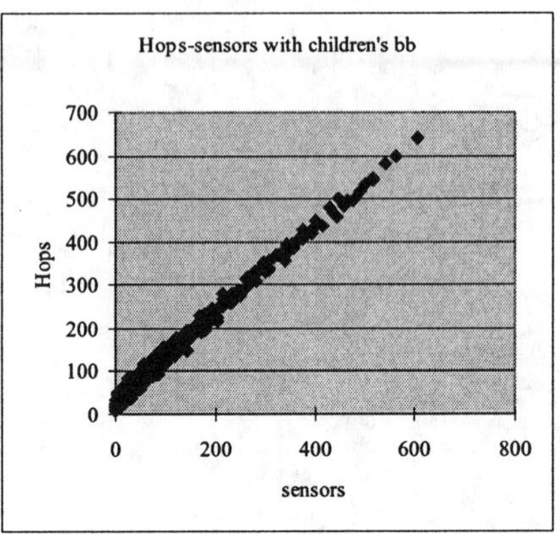

Figure 4: Simulation results.

2.5 Selective Filtering During Broadcasts

In this section, we discuss application of georouting to spatial data broadcasts; in this case, the benefits of georouting apply even when the broadcast is not localized.

When the data being broadcast is a spatial relation, consisting of many spatial features each with its own geographic extent, only a subset of this relation may be relevant to any given sensor node for its computation. When the broadcast is not localized, simple *boolean filtering*, that decides whether to transmit the data to this sensor or not, does not reduce the amount of communication involved in the broadcast. Instead, we can use *selective filtering*, that decides how much of the data to transmit, if any.

To perform selective filtering in georouting trees, we compute the intersection of the sensor's bounding box and the bounding boxes of the spatial features that are candidates for transmission; only those features that intersect the sensor's box are transmitted. This is illustrated in figure 5.

3 SENSOR TERRAINS

In this section, we discuss *delta-gathering*, a technique for reducing communication during data gathering. We then apply our delta-gathering approach toward the problem of sensor data *visualization*. We present *sensor terrains* as an important alternative to isoline-based visualization (*contour maps*).

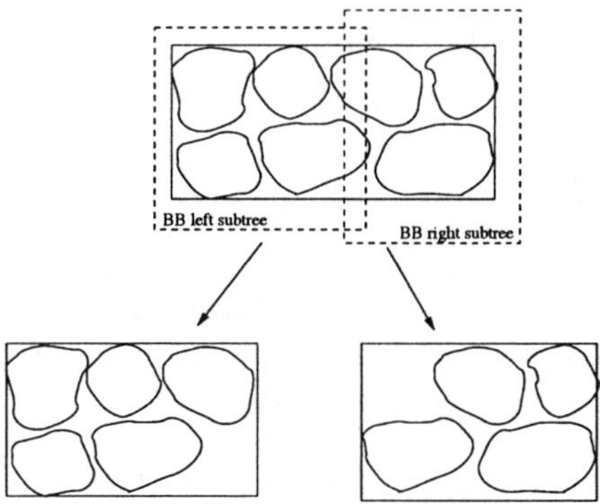

Figure 5: Selective filtering of spatial data in georouting tree.

3.1 Delta-Gathering

For many types of sensor readings, such as *temperature* or *pressure*, there is very little change in value from one epoch to the next. Rather than transmit the readings of all sensors at all times, we only need to transmit readings when there has been sufficient change. In this section, we introduce a new technique to accomplish this, called *delta-gathering*.

Delta-gathering is not to be confused with *delta compression* [23, 29], a related technique. In *delta compression*, we transmit a new value only when the change from the last transmitted value is above some threshold. Delta compression is performed *explicitly*, by specifying the threshold and storing the old value for comparison. This can be done either directly in the query (TinyDB) or with a built-in function (CQL):

TinyDB query with delta compression:

```
SELECT light
FROM buf, sensors
WHERE |s.light - buf.light| > t
OUTPUT INTO buf
SAMPLE PERIOD 1s
```

CQL query with delta compression:

```
SELECT Istream(delta_compr(light))
FROM Sensors
WHERE location = 'NEST-1012'
```

As a result of delta compression, the number of data elements in the stream is reduced. For example, the adjacent values in the output streams of the above queries are guaranteed to differ by more than the tolerance value.

Our new alternate approach, *delta-gathering*, does not involve the difference operator. Instead we are only interested in those values which represent "crossing a threshold".

> **Delta-gathering:**
> Let J be the set of *threshold* values. Let x be the last transmitted value, and y be the current sensor reading; w.l.o.g., assume that $y > x$. y is transmitted only if the interval $(x, y]$ (which excludes x but includes y) contains some value in J.

For example, if the thresholds J consist of multiples of 1, and the latest transmitted value was 2.3, then only the last value in the following sequence will be transmitted: 2.5, 2.7, 2.9, 3.1. Note that $3.1 - 2.3 = 0.8$, which is less than 1.

The goal of delta-gathering is to improve power consumption of the sensor network by reducing the amount of communication at the gathering phase. In the absence of a new value from some sensor, unless we know that the sensor is down, we assume that the value at this sensor has not appreciably changed since the last transmission, and is not worth transmitting.

> Unlike delta compression, this technique does not affect the *semantics* of the data, only the method of gathering.

The data is not compressed; we acknowledge that the untransmitted reading exists and should be part of the data, but we assume that the last transmitted value provides a sufficient substitute for it. This assumption is important for sensor data mining applications such as data visualization, discussed next. When visualizing the data, we will continue displaying the latest known reading for every sensor, until we are notified that it has changed.

3.2 3D Visualization of Sensor Readings

Good visualization of the streaming data produced in sensor networks will enable better monitoring effect of sensitive environmental parameters such as temperature, providing people capacity to respond to alarming changes and make instant decisions. Visualization with *isolines* has been considered in [12]; we have chosen to use *sensor terrains* instead.

We represent a *sensor terrain* as a *triangulated irregular network* (TIN), which is a set of contiguous triangles without overlap. Its vertices are 3D points (x, y, z) where (x, y) is the location of a sensor and z is the reading at that sensor. The TIN representation is popular in *terrain mapping* [7] because of its capacity to represent terrains over irregularly scattered data points, such as the case here.

There are several reasons to prefer sensor terrains to contours as the means of sensor data visualization:

- *more intuitive*: 3D surfaces are cognitively easier than contour maps; for example, differences in height are directly recognizable whereas in isolines, values have to be interpreted
- *less lossy*: we can extract a contour map from the sensor terrain, but not vice-versa
- *greater manipulability*: graphic manipulations of sensor terrains, such as rotations or changes to shading, can further enhance our understanding of the data; this is not possible with isolines
- *easier updates* (for 2D TINs): if one sensor changes value, then only the z-coordinate of that point changes; by contrast the contour map requires more change

An alternative representation to TINs for terrains over irregularly scattered data points is NURBS [27]. This representation is more time consuming to generate and maintain. Another advantage of TINs is the ease of shading, and of extracting isoline information. To be precise, in sensor networks we have a *dynamic* version of TINs and NURBS, where the z values are continuously changing. As the sensor readings change, so does the terrain – it is more like a video than a static surface.

3.3 Dynamic TINs: Overview

There are three basic algorithms for constructing the triangulated representation of a sensor terrain [37]:

- *divide-and-conquer* [10] divides the original data sets into disjoint subsets and solves the subproblem recursively;
- *sweepline* [6] constructs valid Delaunay edges by sweeping the points upward one at a time;
- *greedy insertion* [10] inserts one site at a time into the triangulation and updates the triangulation by iteratively replacing the invalidated edges.

Based on whether the triangulation algorithm makes use of the z values (rather than just x and y), the algorithms are classified as $3D$ (also known as *data-dependent*) or $2D$ (also known as *data-independent*). In the $2D$ case, the triangulation depends only on the sensor locations and not on their readings; in the $3D$ case, it depends on the readings as well.

In the dynamic setting like ours, we assume that the TIN has already been computed, with one of the methods above; instead, we are concerned with *updates* to the TIN. There are three types of updates:

1. *modify value*: corresponds to a change in sensor reading
2. *insert vertex*: corresponds to a sensor joining the network
3. *delete vertex*: corresponds to a sensor leaving the network

The difference between $2D$ and $3D$ TINs is clearest in the case of the first type of update, *modify*; we are assuming *delta-gathering* (section 3.1), so presumably the reading has crossed a threshold. In the $2D$ case, we only need to modify the z attribute of one vertex; the triangulation stays the same. By contrast, in the $3D$ case the triangulation may change.

All updates to the sensor network are placed into an *update queue* at the central processor. They are processed one at a time, to maintain a dynamic TIN whose geometry visualizes the sensor terrain. To maintain the dynamic TIN in real time, two assumptions must be made. First, we assume that the number of updates per epoch is small. This assumption is made feasible by applying *delta-gathering*. Second, we assume that each update is computed very quickly, i.e. with time complexity $O(log\ n)$, where n is the size of the network. In the next section, we discuss the algorithms that make it possible.

3.4 Efficient Updating of TINs

In case of sensor networks, where the updates we must display the surface dynamically and in real time as the updates stream in. Therefore, we found $2D$ triangulation preferable for sensor networks; the triangulation is precomputed and fixed, until a new sensor needs to be added. For adding new sensors, we use the greedy insertion triangulation algorithm.

In this section, we describe the insertion algorithm for the TIN representation of sensor terrains; the *delete* operation is handled in a similar fashion. This algorithm is based on the algorithm for incremental site (vertex) insertion that is part of the *greedy insertion* triangulation algorithm for constructing a $2D$ TIN, found in [10].

Insert. Our *insert* algorithm for $2D$ triangulation closely follows the logic from [10]. Assuming that S is the new vertex to be inserted, it consists of the following steps:

1. **Locate** the triangle T where the vertex S will be located.
2. **Connect** the vertex S with each vertex of the triangle T.
3. **Initialize** the *list of suspect edges* to contain all the edges of T.
4. Remove a *suspect* edge from the list and **test** to determine whether it is *valid*.
5. If invalid, **replace** it with its *alternate*, adding new suspect edges to the list.
6. *Repeat* the last two steps while there are still suspect edges.

In [10], the invalid edges are identified with the *inCircle test*), which dictates that no vertex can be within the circumcircle of any triangle to which it does not belong.

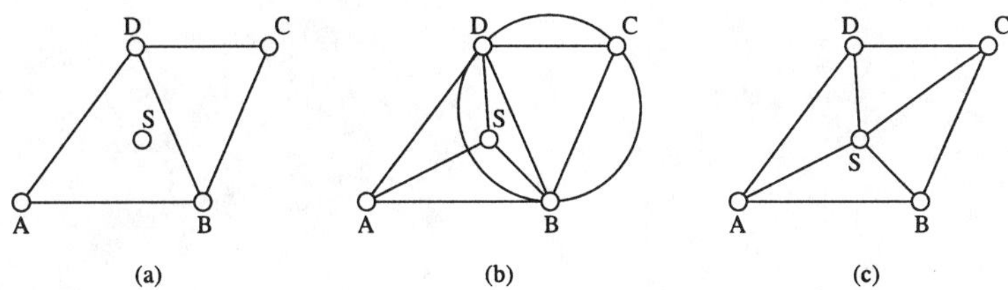

Figure 6: Incremental TIN update in 3 steps

Example. In figure 6 (a), S is the new site to be inserted, and we find that it lies inside the triangle ABD. In figure 6 (b), we connect S to these vertices and run the *inCircle test* for edges AB, AD and BD. We discover that the edge BD is invalid because S is located inside the circumcircle of BCD. In figure 6 (c), BD is replaced by SC. Note that we are not done. Now, BC and CD have become suspect and need to be checked; this procedure is repeated until all invalid edges are removed.

Bounded Change Propagation. As described above, the worst-case performance for *insert* is $O(n)$, due to *change propagation*: all the edges in the triangulation might need to be tested for validity. To ensure $O(log\ n)$ performance, we adapted a *bounded change propagation* strategy: for each update, the maximum number of tested edges is bound at $c\ log\ n$, where c is a constant defined outside our algorithm. With this strategy, the triangulation is no longer *correct* in all cases; hence, the dynamic TIN maintained by our system is *approximate* rather than *exact*. Note that our algorithm is *adaptive*: by increasing c, we can better approximate the correct TIN.

3.5 Simulation of Sensor Terrain Update

We used a sensor terrain of 257 sensors, with coordinates whose x values were randomly distribed in a [0, 9600] range and y values in a [0, 10115] range (this range represented the UConn campus). For our sensor reading, we used actual data for the geographic terrain around the UConn campus, where the sensor readings represent the local height, which is from 0 to 420 feet, when adjusted.

Figure 7 shows the shaded TIN (a) before and (b) after a sensor in the lower left quadrant changed its value, from 350 to 149.49. One can clearly see the difference in the shape of the two sensor terrains.

Figure 7: TIN update example: shaded image (a) before and (b) after update.

4 DYNAMIC ISOLINE EXTRACTION FROM SENSOR TERRAINS

In section 3, we have presented *sensor terrains* as an important alternative to isoline-based visualization (*contour maps*). We have also shown how to maintain a dynamic sensor terrain by incremental updates. In this section, we discuss how to build and maintain a *dynamic contour map* from the dynamic sensor terrain.

We assume that the segments comprising the isolines in the contour map have been computed once from the TIN representing the sensor terrain. Our focus is on *updates* to the TIN, discussed in section 3.3, which necessitate updating the contour map accordingly. The goal is to maintain the TIN and the isolines in real time, for real-time *visualization* of the sensor network. One can imagine the contour map displayed together with the sensor terrain; both of them move on the screen to portray the current state of the sensor network.

For our algorithm, we assume that we can assess the triangles and vertices of the TIN in constant time. We are also assuming *delta-gathering* (section 3.1), so the vertices are only updated when their z value crosses some threshold. It is probably advisable if the set of thresholds for delta-gathering includes the isoline heights of the contour map that is being computed.

We will first present *interval trees*, a data structure that plays a central role in isoline extraction. Given a TIN, the interval tree is computed from this TIN; isoline segments are then computed from the interval tree.

4.1 Interval Trees

Every edge e in a TIN has a *z-span*, which in an interval indicating the minimum and maximum z values in e. Suppose the two end-points of some edge are e_0, e_1, and their height values are r_0, r_1 respectively, where $r_0 \leq r_1$. The z-span for the edge would be $[r_0, r_1]$.

Let Z be the set of all the z-spans of a given TIN. Then, the *interval tree* over this TIN is a binary tree whose nodes are labeled with the following two attributes:

- some *split value* s
- the subset of Z consisting of those intervals that overlap s

Interval trees obey the following properties:

1. Given a node X with split value s, a z-span I of the form (a, b) is in the interval list of X if and only if $a \leq s \leq b$
2. If node Y is a left (right) child of node X, then the split value at Y is smaller (larger) than the split value at X.
3. If the tree has n nodes, then the depth of the tree is $O(log\ n)$.

Our algorithm to extract an interval tree from a TIN is similar to the one in [17]; the major difference is that they have an interval for every *triangle* rather than *edge*. We found edges more convenient for our dynamic implementation.

Figure 8 (a) gives an example of a TIN; figure 8 (b) shows the corresponding interval tree. The lists of intervals are displayed twice, sorted first by start point and then by end.

4.2 Updating the Interval Tree after Change to Sensor Reading

A change to the value of any sensor in the network will affect the triangulation, and hence the set of its z-spans. The interval tree needs to be updated accordingly, so it continues to satisfy the three properties listed in section 4.1.

To update the interval tree, two operations may need to be performed:

1. *update the interval lists*: without changing the *split values* at any of the tree nodes, we modify the interval lists so the first property of interval trees is satisfied
2. *rotate*: without changing the attributes at any nodes, we rotate the interval tree to decrease its height

During the first step above, a new leaf node may have to be added if there are intervals that do not belong to the lists of any of the current nodes. Also, a node will be deleted if its list of intervals is empty.

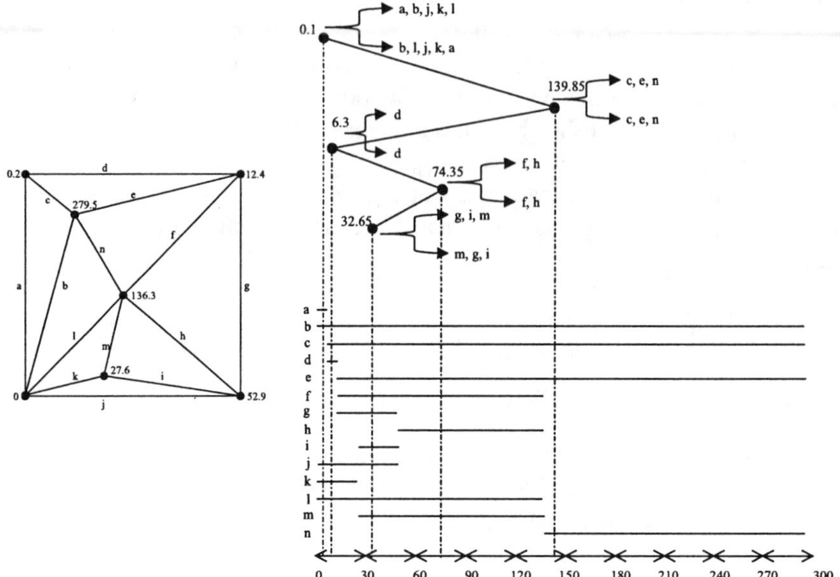

Figure 8: TIN (a) and corresponding tree (b).

Without going into the details of this step, we illustrate it in figure 9, where the sensor reading for the left middle sensor (figure 8 (a)) has changed from 136.3 to 170.4. This figure shows this changes the set of z-spans, and correspondingly the interval tree (before rebalancing). After changing, there are no longer any intervals that lie completely to the left of the root's split value 74.35. There is also a new leaf on the right, whose split value is 224.95. The time complexity of step 1 is $O(log\ n)$, where n is the size of the interval tree.

Clearly, the tree in figure 9 is unbalanced. Figure 10 shows the same tree after a rebalancing (step 2). We use the AVL rebalancing scheme [31] for our interval tree updates, to obtain the overall time complexity of $O(log\ n)$ for our algorithm.

Note that we can *defer* the rebalancing of the tree. That is, we assume that there exists a predetermined constant c such that step 2 is done only once out of every c times that step 1 is done. If the size of the interval tree is initially n, then the time complexity of AVL tree rebalancing after c updates is $O(c(log(c+n)))$ [21].

Figure 11 shows the isolines, computed for the sensor terrain in figure 7 (a), then updated when a sensor in the lower left quadrant was changed from 350 to 149.49. The thick lines represents isoline values of 200 and 300, respectively. The change to the isoline contours is clearly visible.

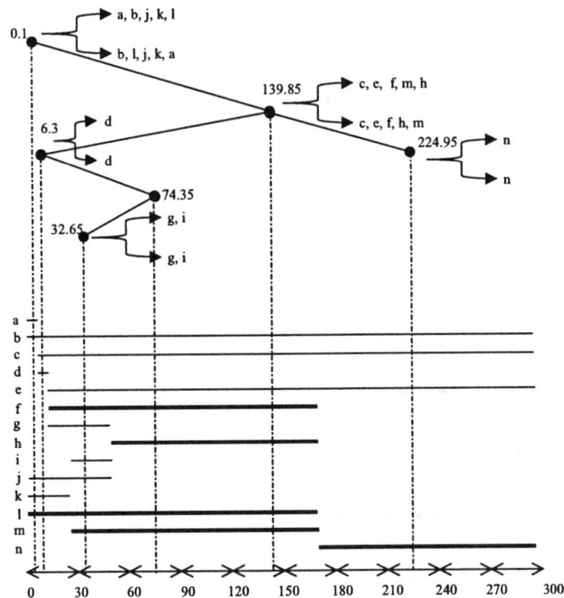

Figure 9: The interval tree after a change of value.

5 CONCLUSION AND FUTURE WORK

We have considered the issue of *query and data propagation* for geosensor network query systems, including our own system SPASEN-SQ. In such systems, techniques for efficient propagation of queries and data play a significant role in reducing energy consumption.

Georouting is a new technique for the broadcasting of *localized data and queries* in geo-aware sensor networks; it makes use of the existing query routing tree, and does not involve the creation of any additional communication channels. In addition to localized broadcasting, georouting is useful for (non-localized) broadcasting *spatial data*, greatly reducing the amount of communication, and hence energy consumption, during broadcasts. We demonstrated its effectiveness empirically, having implemented this technique.

In addition to broadcasting queries and data to the sensors, we considered *data gathering*, where data is being transmitted from the sensors back towards the central processor. *Delta-gathering* is a new technique to reduce the amount of communication during data gathering. We noted that unlike *delta compression*, a related technique, delta-gathering does not affect the *semantics* of the data, only the method of gathering.

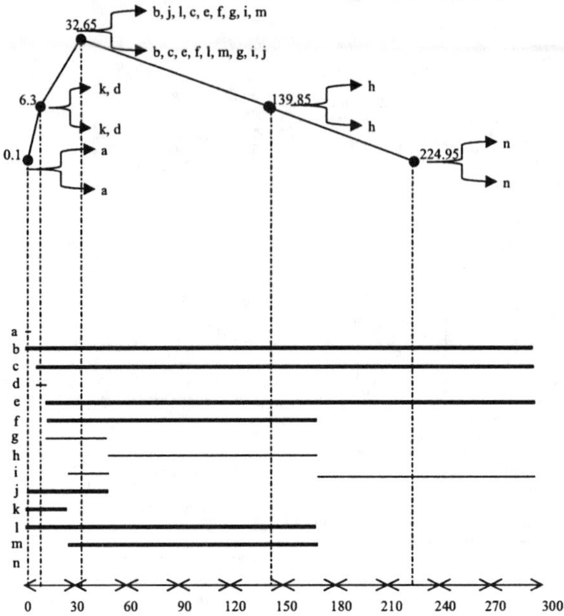

Figure 10: The interval tree after rebalancing.

Finally, we applied delta-gathering toward the problem of *sensor data visualization* via *sensor terrains*. Sensor terrains are a preferable alternative to isoline-based visualization (*contour maps*) for this problem. Sensor terrains are represented by *triangulated irregular networks* (TINs). Visualization of sensor terrains is therefore a special case of *dynamic TIN generation*, a computational geometry problem for which we present a new incremental delta-based algorithm.

Future work includes a real-time interactive sensor terrain and isoline visualization tool which relies on delta-gathering, built into SPASEN-SQ. We also plan to study *in-network* algorithms for the problems discussed above.

References

[1] Braginsky, D. and Estrin, D., Rumor Routing Algorithm For Sensor Networks, *In Proc. First ACM Int'l Workshop on Sensor Networks and Applications (WSNA)*, Atlanta, GA, Sep. 2002.

[2] Bertino, E., Guerrini, G., and Merlo, I., Trigger Inheritance and Overriding in an Active Object Database System, *IEEE Transactions on Knowledge and Data Engineering*, 12:4, pp. 588–608, 2000.

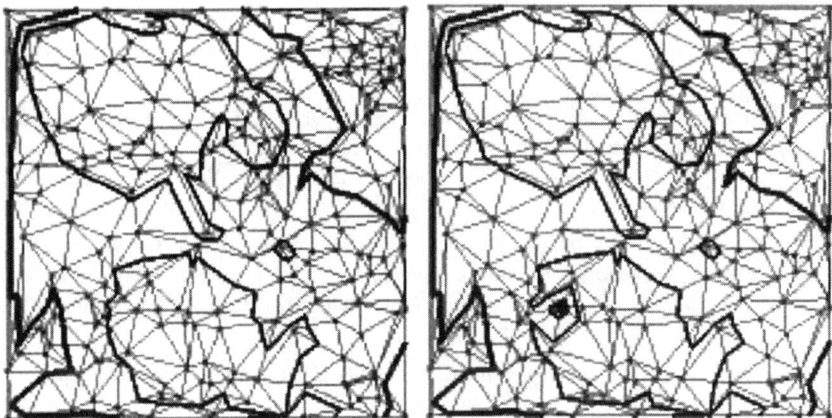

Figure 11: TIN update example: isolines (a) before and (b) after update

[3] Cerpa, A. et al., Habitat Monitoring: Application Driver for Wireless Communications technology, *ACM SIGCOMM Workshop on Data Communications in Latin America and the Caribbean*, Costa Rica, April 2001.

[4] Chang, J-H. and Tassiulas, L., Energy Conserving Routing in Wireless Ad-hoc Networks, in *Proc. IEEE Infocom*, pp. 22-31, Tel Aviv, Israel, March 2000.

[5] Elmasri, R. and Navathe, S., *Fundamentals of Database Systems*. Addison-Wesley, New York, 2000.

[6] Fortune, S., A Sweepline Algorithm for Voronoi Diagrams, *Algorithmica*, 2:153-174, 1987.

[7] De Floriani, L., Puppo, E., and Magillo, P., Applications of Computational Geometry to Geographical Information Systems, Chapter 7 in *Handbook of Computational Geometry*, J.R. Sack, J. Urrutia (Eds.), Elsevier Science, pp.333-388, 1999.

[8] Garland, M. and Heckbert, P.S., Fast Polygonal Approximation of Terrains and Height Fields, Technical Report CMU-CS-95-181, Carnegie Mellon University, 1995.

[9] Gehani, N. and Jagadish, H.V., Ode as an Active Database: Constraints and Triggers, *Proc. 17th Int'l Conference on Very Large Databases*, 1991.

[10] Guibas, L. and Stolfi, J., Primitives for the Manipulation of General Subdivisions and the Computation of Voronoi Diagrams, *ACM Transactions on Graphics*, 4(2):75-123, 1985.

[11] Heidemann, J. and Bulusu, N., Using Geospatial Information in Sensor Networks, in *Proceedings of the Computer Sciences and Telecommunications Board (CSTB) Workshop on the Intersection of Geospatial Information and Information Technology*, Arlington, VA. October, 2001.

[12] Hellerstein, J.M. et al., Beyond Average: Towards Sophisticated Sensing with Queries, *2nd Int'l Workshop on Information Processing in Sensor Networks (IPSN '03)*, March 2003.

[13] Intanagonwiwat, C., Govindan, R., and Estrin, D., Directed Diffusion: A Scalable and Robust Communication Paradigm for Sensor Networks, In *Proc. Sixth Annual International Conference on Mobile Computing and Networks*, August 2000, Boston, MA.

[14] Karp, B. and Kung, H.T., Greedy Perimeter Stateless Routing for Wireless Networks, in *Proceedings of the Sixth Annual ACM/IEEE International Conference on Mobile Computing and Networking (MobiCom2000)*, Boston, MA, August 2000, pp. 243-254.

[15] Kuper, G., Libkin, L. and Paredaens, J. (Eds.), *Constraint Databases*. Springer-Verlag, Heidelberg, 2000.

[16] Kulik, J., Rabiner, W., and Balakrishnan, H., Adaptive Protocols for Information Dissemination in Wireless Sensor Networks, *Proc. 5th Int'l Conf. on Mobile Computing and Networking*, Seattle, WA, 1999.

[17] Van Kreveld, M., Efficient Methods for Isoline Extraction from a Digital Elevation Model Based on Triangulated Irregular Networks, in *Proc. Sixth Int'l Symposium on Spatial Data Handling*, pp.835-847, 1994.

[18] Kung, Vlah. Efficient Location Tracking using Sensor networks, *Proc. 2003 IEEE Wireless Communications and Networking Conference*.

[19] Leach, G., Improving Worst-Case Optimal Delaunay Triangulation Algorithms, in *Proc. 4th Canadian Conference on Computational Geometry*, 1992.

[20] Li, Q. et al., Reactive Behavior in Self-Reconfiguring Sensor Networks, *ACM MobiCom 2002*, September 2002.

[21] Larsen, K.S., Soisalon-Soininen, E., and Widmayer, P., Relaxed Balance through Standard Rotations, Workshop on Algorithms and Data Structures, 1997

[22] Madden, S.R. et al., *TAG: a Tiny AGgregation Service for Ad-Hoc Sensor Networks, OSDI*, December 2002.

[23] Madden, S.R. et al., The Design of an Acquisitional Query Processor for Sensor Networks, *SIGMOD*, June 2003, San Diego, CA.

[24] McErlean, D. and Narayanan, S., Distributed Detection and Tracking in Sensor Networks, *36th Asilomar Conference on Signals, Systems and Computers*, 2002.

[25] Madden, S.R. et al., Supporting Aggregate Queries Over Ad-Hoc Wireless Sensor Networks, *Workshop on Mobile Computing and Systems Applications,* 2002.

[26] Rigaux, P., Scholl, M., and Voisard, A., *Spatial Databases,* Morgan Kaufmann, 2001

[27] Song, M., Goldin, D.Q, and Peng, T., NURBS Surface Interpolation for Terrain Modeling, to appear in *Proceedings of ASPRS/MAPPS 2003 Conference on Terrain Data*, October 2003, North Charleston, SC.

[28] Silberschatz, A., Korth,H., and Sudarshan, S., *Database System Concepts*. McGraw-Hill, New York, 2002.

[29] The STREAM Query Repository, Stanford University.

[30] Woo, A. and Culler, D.E., A Transmission Control Scheme for Media Access in Sensor Networks, *Proc. 7th Int'l Conf. on Mobile Computing and Networking*, Rome, Italy, July 2001.

[31] Weiss, M.A., *Data Structures and Algorithm Analysis in C*, Addison-Wesley, 1997.

[32] Adjue-Winoto, W. et al., The Design and Implementation of an Intentional Naming System, in *ACM SOSP*, December 1999.

[33] Xu, Y. and Heidemann, J., Geography-informed Energy Conservation for Ad Hoc Routing, *Proc. 7th Int'l Conf. on Mobile Computing and Networking*, Rome, Italy, July 2001.

[34] Yao, Y. and Gehrke, J., The Cougar Approach to In-Network Query Processing in Sensor Networks, In *SIGMOD Record*, September 2002

[35] Yao, Y. and Gehrke, J., Query Processing in Sensor Networks, *CIDR 2003*, January 2003.

[36] Yu, Y., Govindan, R. and Estrin, D., Geographical and Energy Aware Routing: A Recursive Data Dissemination Protocol for Wireless Sensor Networks, UCLA Technical Report UCLA/CSD-TR-01-0023, May 2001.

[37] Su, P. Efficient Parallel Algorithms for Closest Point Problems, Ph.D. Thesis, Dartmouth College, NH, 1994.

Information Handling in Mobile Applications: A Look beyond Classical Approaches

Jochen Schiller

Freie Universität Berlin
schiller@pcpool.mi.fu-berlin.de
Germany

Agnès Voisard

Fraunhofer ISST Berlin,
and Freie Universität Berlin
agnes.voisard@isst.fraunhofer.de
Germany

ABSTRACT

Mobile devices such as cellular phones or personal digital assistants now allow end users to obtain information based on their current location. This information is usually based on distance (e.g., the nearest drugstore) and navigation instructions. Richer information may also be delivered to mobile users according to their profiles, for instance, their personal preferences, and to their history, i.e., in relation with the information that they have already received. A new dimension can be introduced if the communication among mobile peers is enabled. Mobile users are then able to exchange information among each other. This paper focuses on two major ways of considering information delivery in this context. The first one is based on a common, centralized approach and considers many users taken individually. The second one is a decentralized approach that considers many users who exchange information among each other. We explain the two paradigms and describe the current possibilities for the underlying infrastructure.

Keywords: location-based services, mobile and wireless ad-hoc networks, peer-to-peer networking, push/pull services, event notification systems

1. INTRODUCTION

Mobile devices such as cellular phones or personal digital assistants now allow end users to obtain information based on their current location, such as the nearest drugstore and even the fastest route to it from the current location. However, in such location-based services (LBS), it is desirable to obtain richer and more targeted information than the pure data stored in a database. Typical mobile objects are cars, boats, or people. A complex mobile application is that of tourists walking around in a city and receiving the right information at the right time. Because of the richness of such applications, we chose them to illustrate our discourse in this article.

The tourists are equipped with a mobile device, such as a personal digital assistant or a cellular phone. They can ask explicitly for information such as information services on the world-wide Web or in traditional location-based services (e.g., nearest coffee shop computed from the location and from yellow pages) but also obtain information related to a point of interest (POI) situated at their location. This is done by referring to such a point, either by pointing to it with a device or by describing it (using its address or a succinct description). Information will be pulled from the server and delivered to the end-user on his/her device.

When tourists walk around, a novel feature is to notify them about relevant events or places to visit in the near future according to their profile (push mechanism). This procedure goes beyond common broadcasting as the whole context of the user - which includes his/her profile - needs to be taken into account: current location, time, interests, and preferences. This information is stored in one or many databases. The interests and preferences of the end-users are entered by subscribing to topics of interest (e.g., Architecture of the XIXth Century) or may be inferred by the system after examining the behavior of the users and grouping user into clusters through data mining techniques.

In addition to receiving appropriate information with respect to their profile, users do not wish to get the same information more than once unless they explicitly ask for it. Hence, the system should have a memory of what has already been delivered to its users. Last but not least, end-users would like in general to establish correlations with information that they received previously, for instance the fact that the architect of the building that he or she is visiting also built another well-known building in the city and that furthermore the mobile user in question has seen that building the day before.

The above examples consider users taken individually, even though a large number of users may be considered. Let us now consider a network of users exchanging information directly among each other to get information on points of interest or events of interest. This can be achieved by sending a message to peers located around, such as "Did someone see the exhibition? Is it worth the long waiting queue?" Then a peer will take the question and answer directly. This mechanism is built on the Peer-to-Peer (P2P) network paradigm.

To be realized successfully, such applications need to rely on many different techniques and concepts that range from service chaining paradigms to wireless technologies such as local area networks (LAN) or Bluetooth. In this paper, we describe the major issues emerging from such applications, both at a conceptual and at a technical level.

With the emergence of techniques to locate users and then of location-based services, a new category of applications referred to as *context-aware* applications has received a great deal of attention in the past years. However, the vocabulary used in such applications is rich and often not "standard".

Terms such as situation, context, surrounding are often used as synonyms without a clear definition. The goal of this paper is not to propose a definition of these terms, however, it is important to keep in mind that we need to consider the following concepts: the current time, the location of the users, their history (where they have been and what information they received), and, last but not least, their profile. User profile is a complex notion. It encompasses the notion of personal preferences from a single user and of a general collection of preferences based on user clusters as described above. Besides, preferences may change according to a location and to a certain time.

Several system exist that process context-aware and location-based information for mobile human users. In the area of tourism, because of technical issues, two major approaches are usually distinguished: services for outdoor experiences and services focusing on indoor activities. As described in [1], examples of outdoor services include tourist guides, such as Nexus [2], Guide [3], Crumpet [4], the Oldenburg guide [5], and CATIS [6]. Examples of indoor systems are museum guides, e.g., Hippie [7], Electronic Guidebook [8], and Rememberer [9]. Database modeling and querying of moving objects is of prime interest in this context – see for instance [10-14] distinguishes three types of queries: instantaneous (evaluated once at definition), continuous (evaluated regularly after definition), and persistent (sequence of instantaneous queries evaluated at every database update). In the context of location based services, continuous queries are of particular interest. Most systems only consider queries regarding the changes in user location. Besides, in most systems, the context is merely the user location measured either at certain access points (e.g., in Electronic Guidebook) or at a given location (e.g., in Nexus). That is, continuous queries, or profiles, only consider the user's location. Additional information such as user interest, local time, or technical communication aspects are often not used.

It is worth noting that only a few systems encompass the notion of events or profiles. In the Guide system, tourists are informed about general events such as the opening hours of a museum. This information is broadcasted to all users of the system and each available sight (whether or not they are interested). In VIT, keyword profiles are used to select the information sources by topic (similar to advertising). In Crumpet, the information delivered on request is sorted according to user interests. The user's interest is defined in terms of profiles that are gradually adapted to the user based on specific user feedback. The Oldenburg guide uses continuous queries regarding the spatio-temporal location of the users as profiles. Information about (moving) objects is delivered depending on its importance to the user, where importance depends on the spatial distance between object and user.

This paper is organized as follows. Section 2 gives our example scenario that serves as a reference throughout the paper in order to illustrate the concepts we introduce. Section 3 is devoted to information delivery to individual users; that is, the server sends information to users' mobile devices. Section 4 focuses on information exchange among end-users. We describe the conceptual and technical underlying issues with a focus on new perspectives in P2P networks. Finally, Section 5 draws our conclusions.

2. EXAMPLE SCENARIO: ENHANCED TOURIST APPLICATION

Let us consider two friends, Carmen and Juliet, walking in Berlin on October 28, 2003. Carmen is interested in Jewish history. Her natural language is Spanish. Juliet is American and is interested in wall-related facts. They speak English with each other. They both like modern architecture and paintings, especially paintings from the GDR. They plan their journey independently but would like to meet around noon for lunch. Their trajectory is as follows.

Carmen walks between the following places:
10:00 Synagogue
11:00 *Reichstag* (Parliament)
11:15 Brandenburg Gate
11:30 *Potsdamer Platz.*
15:00 *Cinemaxx*
15:15 *Neue Nationalgalerie*

Juliet walks between the following places:
10:00 *Unter den Linden*
11:00 Brandenburg Gate
11:45 *Potsdamer Platz*
11:50 *Marlene Dietrich Platz*
15:00 *Cinemaxx*
15:15 *Neue Nationalgalerie*

At the Synagogue, Carmen is given information on her PDA regarding its architect and opening hours. She then walks toward the Reichstag. The system tells her that she is crossing the former wall. She continues walking. She arrives near the Reichstag and sees a glass cupola sticking out. She would like to find out what this monument is so she sends a request to the system by pointing to this monument and entering the keyword "cupola". The system gives her information on the Reichstag. She then goes toward the Brandenburg Gate where she is told that a European Jewish Memorial is about to be built and that its construction is controversial. She is offered access to the articles relating the latest developments, both in Spanish and in English.

During that time, Juliet is walking from the *Unter den Linden* Avenue to the Brandenburg Gate, where the system tells her that she is at the place where the wall used to be. She continues walking. The system tells her every once in a while that she is going along the former wall and that she could walk toward the east for 10 minutes and see some of the remaining wall. The system offers her pictures. She then asks the system where she can go shopping. The system gives her information on the shopping mall at *Potsdamer Platz*. She heads down toward that place and calls her friend Carmen to synchronize for lunch. She asks the system to find the most appropriate place for lunch given their respective locations and their medium budget.

At 3 pm, as they walk around *Postdamer Platz* they are notified that *Ice Cream Paradise* has happy hour in 10 minutes. They are given the directions to go to the place, stay there, and decide to go to the movies. They do not know what movie to pick. They hesitate between Movie A *"Matrix - Revolutions"* and Movie B *"Kill Bill – Volume I"*. They send a message to people who are exiting out the movie theater and ask who has seen A and/or B. Many people answer that B is not a movie to miss so they decide to see its next showing. The system tells Juliet that she has seen another Quentin Tarantino movie the week before in Paris.

When they get out of the movie theater and walk toward the casino, they are both notified, Carmen in Spanish and Juliet in English, that there is an exhibition at the *Neue Nationalgalerie* with the theme "Art during the GDR", which is open until 10 pm tonight. They ask for the time it takes to go there, as well as the approximate time of the queue. Then they decide to go there and to book their ticket online.

3. INFORMATION DELIVERY TO INDIVIDUAL USERS

In the examples above, the end-users are considered as collections of mobile users who explicitly ask for information and who are sent information with the assumption that it is of interest to them. This is done by combining the two paradigms of location-based services and event notification systems and is referred to as a centralized approach. A second approach consists in sending any kind of information – relevant or not - to the device of the mobile user, which then filters the information for the user in question according to his or her profile. This *geocasting* approach is not detailed in this paper.

3.1 Combining location-based services and event notification systems

The main terms used in the context of event notification systems (ENS) are *events* and *profiles*. An *event* is the occurrence of a state transition at a certain point in time. In [15] we make the distinction between primitive and composite events. A primitive event describes a single occurrence, while a composite event describes the temporal combination of events (e.g., a sequence). Composite events can be described using an event algebra. Composite event operators are, for instance, sequence $(E_1;E_2)_t$, conjunction of events $(E_1;E_2)_t$, disjunction $(E_1;E_2)_t$, and negation $(E_1)_t$. For example, a *sequence* $(E_1;E_2)_t$ occurs when first $e_1 \in E_1$ and then $e_2 \in E_2$ occurs. The parameter t defines in the profile the maximal temporal distance between the events.

A *profile* can be seen as a query executed on the incoming events. In ENS, the result of an event that evaluates successfully against a certain profile is a notification about the event. Unmatched events are not considered. Besides, in mobile applications, we distinguish the following types of events:

- *Location events* are events that are connected to a specific user, time, and location. A location event occurs when the user presses the *Information*-button.
- *External events* are caused by external sources that send event messages to the system. External events are also connected to a certain time, location, and profile and are pushed to the concerned users. Location events trigger a system reaction that results in the dissemination of information to the respective users. External events are, depending on the users' profile and location, forwarded to selected users.

In an ENS, the action triggered by an event is the possible forwarding of the event information. In our system, the following three forms of actions are distinguished, all based on a push mechanism:

1. *Information Delivery*. In this case, the action defined in the profile specifies the information data to be selected from the database and to be sent to the user. The selected information data depends on the location event, its time and location, on the event history, on the user/generic profiles, and on the semantic network of the information data. Depending on personal profiles, only selected information about a certain sight - or attraction - is delivered. Depending on generic profiles, additional information may be delivered about the interconnection of sights already seen. An important type of notification is the spatial notification, where the system establishes a correlation between the trajectory of the user and the geometry of a point of interest using customized spatial predicates such as "nearby" or "less than 5 meters than".
2. *Recommendations*. Here, additional information about semantically-related items is given. The selected information depends on the information event, its time and location, the history of events, the user profile, and, last but not least, the semantic network of information data (with a notion of clusters).
3. *Scheduled/ External Message Delivery*. In this form of action, the delivery depends on the external/scheduled event, the time and location it refers to, and the user profile.

In our system, the profiles are similar to triggers in active databases or profiles in an event action system. In contrast to profiles in ENS, the profile structure is not defined as event-condition (-notification) but as event-condition-action. The action is defined as the selection of information from the various databases. This information is not extracted from the event message but from the system databases. As far as the mobile user is concerned, we distinguish the following two kinds of profile:

1. *Personal profiles* are associated with end users. They are either defined explicitly by the end user or inferred from user actions applying user profiling techniques. The personal profile influences the information selected for the user. An example of a personal profile is "Send only information about architectural facts". Simple personal profiles consist of keywords selecting topics of information. More advanced personal profiles may consist of attribute-value pairs or database queries that specify certain information. For example, the recommendation of restaurants may be based on user profiles defining the food style (e.g., Italian), the price level (e.g., moderate), and additional restrictions (e.g., vegetarian).
2. *Generic profiles* are defined in the service. They are based on a general structural relation between the information data. An example of a generic profile is "Advise the visit of all monuments that are in the same semantic group as the one visited and have not been visited yet, provided that the user has already seen 70% of them". Simple generic profiles may use only the most recent location event, while sophisticated generic profiles are based on user event histories.

At another level, application profiles are application-specific profiles defined by an application expert, e.g., the provider of the tourist information. For example, a tourist information guide provides specific information to be delivered if the tourist visits a certain sight after another one. This mechanism relies on the history of the user.

Let us get back to recommendations and to the notion of delivering the right information at the right time according to the profile and to the history, which is a quite complex action. The system notifies the mobile user of a supposedly interesting piece of information. This is based on the observation that the user has seen many "similar" sights. Then other similar sights are recommended. The question is how to recognize similar sights. A trivial way is to consider classes of sights, such as all cathedrals in Berlin. A more elaborate way is to link sights together if they have something in common - for instance, the same architect. An even more sophisticated way, which relies on mining techniques, consists in making associations with what mobile users have already seen, even though some sights may *a priori* have nothing in common, and to recommend such sights to the users provided that they have not been visited yet and that they are nearby.

3.2 Back to the scenario

The simple scenario described in Section 2 illustrates many concepts of information delivery. In the following let us concentrate on information delivery to individual users. First, the two users have their respective profiles, which encompass personal data as well as preferences. In database jargon we refer to this personal information as *attributes*. Attributes range from static, e.g., names and first names ("Carmen" and "Juliet" here), to highly dynamic, e.g., the location of the user which changes all the time. Other attributes concern the preferences of the user in general or at a certain time. For instance, Carmen's natural language is Spanish but in some situations - e.g., when talking to her friend Juliet - she prefers English.

Besides, users can subscribe to topics of interest. Carmen subscribes to the topic "Jewish history" and Juliet to "Wall-related facts". They both subscribe to topics "architecture" and "GDR painting". This is typically done on a personal computer when preparing a trip or "on the fly" on the mobile device. When the system finds interesting information that matches the topic, e.g., an exhibition on GDR painting nearby, it notifies the users and sends recommendations on an Event of Interest (EOI). In this example, the recommendations are based on subscription topics, time, and location. Another type of location is based on a more sophisticated operation, which consists of comparing the trajectory of the mobile user with the geometry of an object of interest, such as the trajectory of Juliet and the wall in the example.

3.3 Architecture of a Centralized Tourism Information Provider

In the centralized approach, such as the Tourism Information Provider (TIP) under development at FU Berlin, the system disseminates information based on time, location, and profiles. More precisely its components are:

- *Mobile devices*. The application scenario described in the previous section illustrates the need to send a location at any time and to ask basic queries. A critical issue is the visibility of the history. For privacy reasons, the history should be stored on the device. With a central approach, this means that each time end users pose a query their history should be shipped to the system. It is up to the user to make parts of the history visible. In other words, location/time events can be removed from the history (e.g., the end user may want to hide the location where he or she has been).
- *Server*. The system hinges on three thematic databases, which are:
  1. *Profile database*. This database contains generic profiles as well as personal profiles.
  2. *Scheduled event database*. This database contains events of interest (EOIs), i.e., events that have a limited validity in time such as programs (e.g., concert schedules).
  3. *Spatial database*. This database contains maps as well as Points Of Interests (POI) such as museums, restaurants - all classified through categories - or teller machines. They are organized in a semantic

network. POIs are gathered in semantic clusters. Note that external events are not stored in a database but are coming from an external source.

4. *Location engine.* It maps locations to maps themselves and assists users in geospatial-related operations. The basic operations are:
   - *Geocoding.* Through this operation, the end user gives an address and the system returns a (longitude, latitude) pair, which may be used to find places of interest in a certain area.
   - *Reverse geocoding.* Through this operation, which is mostly used here, the user sends periodically a (longitude, latitude) pair and the system returns an address.
   - *Routing.* As seen from the typical LBS query in the example (i.e., *Where can I find a shopping area nearby?*) we need navigation tools. The algorithms usually use a two-sided (memory-bounded) A* algorithm to route someone from one coordinate to another (note that geocoding/reverse geocoding are often coupled with this algorithm).
   - *Proximity search.* This is a broad issue as it concerns many dimensions: time, location, and semantics. The buffer allowed in each dimension can be set by default depending on the profile of the user (e.g., when walking, "nearby" means about 2 minutes for this particular tourist, when considering distances, 200 meters) and may be changed. With the spatial database, it is a typical point query or region query with a buffer of e.g., 200 meters around the location (plus fuzziness).

The notification system does the following tasks:

- Compares the profile of the user to deliver relevant information.
- Looks for relevant information in the scheduled events and spatial databases by applying spatio-temporal operators.
- Looks for external events.
- Processes typical LBS queries.
- Compares the situation with the *profile* of the user and the relevant part of his/her *history* to deliver information.

End users define profiles regarding certain items and topics, e.g., architecture. In the event history, for each end user, the events are stored together with their location and occurrence time. We assume that an end user has a location at a certain time, which is a 0-dimensional entity in a 2-dimensional space. The location associated with an object of interest is a (two-dimensional) area. The data model used in the TIP system relies on RDF and is described in [1].

4. INFORMATION EXCHANGE AMONG MANY USERS

This section focuses on the case where many mobile users exchange information, such as in the second part of the scenario of Section 2. These users form implicit groups that are based either on location (for instance, all the people who are in a 500 meter radius from Carmen) or on profiles (e.g., all users interested in GDR painting). The paradigm of mobile peer-to-peer (P2P) networks fits perfectly the requirements of our scenario: groups of mobile users with similar interests want to exchange information without access to a fixed infrastructure, content providers, or any centralized database. In this section, before we present our approach for mobile networks [16] and give a comparison with other approaches, we introduce P2P networks in general and discuss the most prominent drawback in more detail: the mismatch of overlay and underlay routing.

4.1 Mobile Peer-to-Peer (P2P)

In contrast to classical client-server networks, P2P networks do not require centralized control or storage of information. In order to participate, a node only has to know a single entry point into the P2P network. Most P2P networks are based on the Internet and establish a so-called overlay network for information exchange. Classical Internet routing protocols still perform routing of IP data packets in the underlying network. Typically, the structure of the overlay network is completely independent of the underlying network topology.

P2P systems have recently seen a tremendous surge in popularity which has led to the development of a variety of such systems. However, first-generation systems such as Gnutella [17] suffer from serious scalability problems [18]. Thus, current research efforts have been devoted to distributed hash tables (DHTs) to overcome these scalability obstacles.

DHTs are self-organizing overlay networks especially tailored toward the need of large-scale peer-to-peer systems. The general idea of DHTs is that each node participating in the (overlay) network is assigned a random ID. Each object that is to be stored on the network is also assigned a random ID. An object is now stored on the node whose ID is closest to the object's ID. All DHTs provide one basic operation: lookup(key) → node. Given an object's ID, a DHT is capable of locating the responsible node within a bounded amount of overlay routing steps. Prominent representatives of DHTs are CAN, Chord, Pastry, and Tapestry [19-22].

In the overlay network, a node maintains an overlay routing table containing the IDs of a small set of other overlay nodes. Each such entry can be thought of as a virtual, direct link between the current node and the table entry. In overlay terms that means that messages can be exchanged directly between a node and the nodes in its routing table or, in other words, a node can reach all nodes in its routing table with a single overlay hop.

However, a single overlay h7op is likely to involve multiple physical routing hops. For example, consider two overlay nodes A and B connected to the Internet. A is located in London and B in Los Angeles. It is quite obvious that even if B resides in A's overlay routing table, the one overlay hop between A and B would amount to several IP hops.

The main advantage of DHTs is that they provide a guaranteed bound on the number of overlay routing hops that have to be taken to locate any given object (i.e. any given key) on the overlay network. For [20-22] this bound is O(log N), where N is the number of nodes participating in the overlay network.[1] Due to the discrepancy between overlay hops and physical hops, as explained above, it is very likely that a significantly larger amount of physical hops compared to the logarithmic amount of overlay hops is involved in locating an object on the overlay network. With high probability, the following issue arises from this discrepancy. The number of physical hops induced by the overlay routing process can be decidedly greater than the direct physical routing path between the source node and the target node.

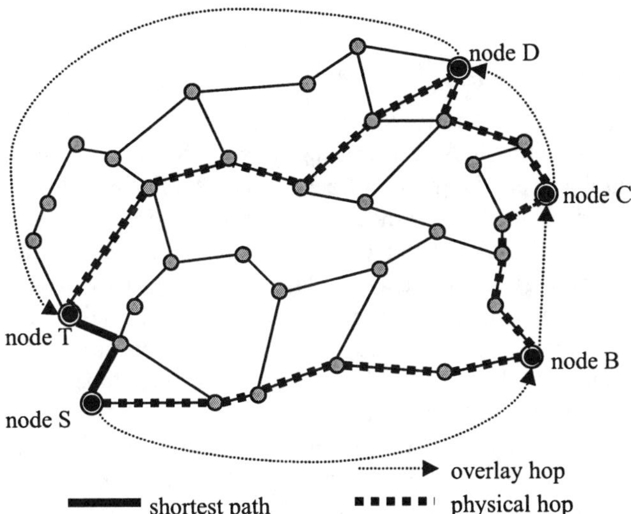

Figure 1: Overlay routing.

[1] CAN employs a more general approach involving d-dimensional virtual coordinate spaces. CAN has a routing effort of $O(d(n^{1/d}))$. However, if the number of dimensions d is chosen to be $d=(\log_2 n)/2$, CAN achieves the same effort as Chord, Pastry and Tapestry.

Consider the overlay routing example given in Figure 1. Overlay node S initiates a lookup that will eventually be routed to overlay node T. Since every overlay node only has very limited knowledge of other overlay nodes, nodes usually try to forward a lookup request to other nodes that are closer (in terms of the overlay ID space) to the key than they are themselves.[2] In this example, three intermediate overlay routing steps are involved until the request reaches its final destination, clearly traveling a highly suboptimal physical route.

As can be seen, although the target node can be located with logarithmic overlay hops, the physical path traveled during the overlay routing process is often less than optimal. More technically speaking, the ratio between the number of physical hops induced by overlay routing and the number of physical hops on a direct physical routing path is often markedly lopsided.

Mobile P2P networks introduce yet another difficulty as now all peers are mobile and may appear and disappear quite frequently. While substantial research results exist for general P2P networks, the area of mobile P2P networks is still widely unexplored. However, P2P and independent, mobile nodes seem to be the perfect match.

4.2 Back to Our Scenario

In section 2, Carmen and Juliet used their PDAs to get opinions regarding two alternative movies that looked interesting to them. Sending a request to other peers (i.e. in this scenario, groups of people with similar interests, interesting knowledge and so on) in an ad-hoc fashion using a wireless mobile network without a fixed infrastructure involves many difficult questions of message forwarding (routing). On the network layer several goals exist for routing: route stability, efficiency, minimum number of hops to reduce latency and save energy etc. [16]. Most of today's research concentrates on the network layer if thinking of ad-hoc networks, that little has been done combining application requirements and network topologies.

Why is the ratio between the number of physical hops induced by overlay routing and the number of physical hops on a direct physical routing path very important in our scenario? Think of Carmen and Juliet using their PDAs for all information access. These devices, as almost all mobile devices, suffer from severe power restrictions and limited stand-by times due to their batteries. If the physical routing is inefficient, mobile devices have to forward many messages in P2P networks using non-optimal paths. This automatically

[2] Exactly how many nodes an overlay node knows about and how message forwarding is done, is an implementation-specific detail of the respective DHT.

involves many more nodes in the forwarding process as required considering the optimal physical path. Each forwarding of a message drains the battery as sending requires a lot of energy compared to stand-by. Without optimized routing paths, Carmen's and Juliet's PDAs would experience much shorter battery life-times.

4.3 The DynaMO Approach at FU Berlin

DynaMO (Dynamic Mobility-Aware Overlays) actively exploits physical proximity in the creation of overlay networks which is imperative in our mobile scenario. DynaMO's main focus is on achieving good locality properties in the overlay network by inducing as little construction and maintenance overhead as possible while still maintaining an even overlay ID distribution. This will translate into an optimized overlay vs. physical routing distance ratio, which is particularly crucial in dynamic networks. Maintaining an even overlay ID distribution is especially important in ad-hoc networks with extremely heterogeneous devices where devices with scarce resources should not become hotspots (compare mobile phones with PDAs, laptops, cars, etc.).

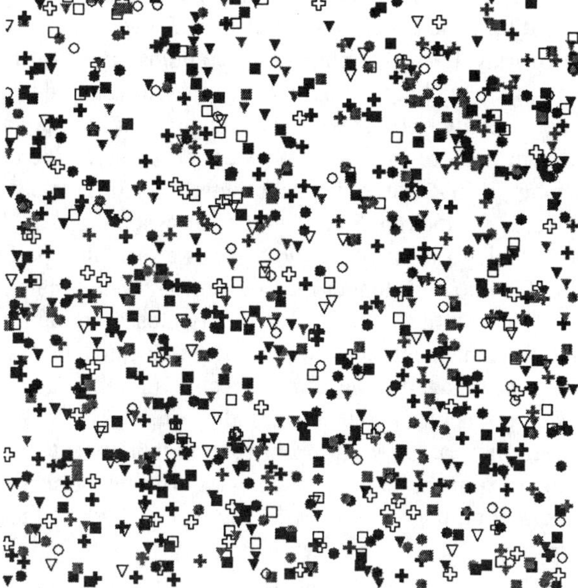

Figure 2: Spatial prefix distribution as generated by Pastry. Equal symbols represent equal prefixes.

Our implementation of DynaMO is based on a Pastry overlay network. Pastry is a well-known DHT that provides built-in locality heuristics. We chose Pastry because these heuristics have been thoroughly analyzed [23].

This analysis makes a good background against which to compare the results achieved with DynaMO. However, we believe that DynaMO's mechanisms are DHT-independent and could, thus, be ported to other DHTs.

DynaMO's approaches differ primarily from Pastry's approach by the way in which overlay IDs are assigned. Pastry's overlay construction basically works in a top-down fashion, i.e. Pastry randomly assigns overlay IDs regardless of the underlying topology. This is reflected in the total randomness of the spatial overlay ID distribution, as depicted in Figure 2. It, then, tries to make the physical proximity fit into the overlay routing state through the join process and table maintenance. In contrast, DynaMO constructs the overlay network in a bottom-up fashion, i.e. the overlay is built considering locality information from the underlying network. Before a node joins the overlay, it gathers information concerning its physical neighborhood and uses it to assign itself an appropriate overlay ID. The most promising approach examined in this context is *random landmarking* (RLM).

To analyze the different effects of Pastry's and DynaMO's approaches, at this point Pastry's overlay routing is discussed briefly (for a thorough discussion see [21, 23]). Each Pastry node essentially maintains a routing table and a leaf set. The routing table consists of a number of rows equal to the number of digits in an overlay ID and a number of columns equal to the ID base. From row to row, the matching prefix between the current node's ID and the row's entries increases by one. The leaf set contains the numerically closest nodes to the current node regardless of physical proximity. When a node has to forward a lookup, it first checks whether the requested ID is covered by its leaf set and forwards the lookup directly to the corresponding leaf. Otherwise, it uses its routing table to identify a node that has a matching prefix with the requested key that is one digit longer than the current node's matching prefix. This process continues until the node numerically closest to the requested ID is located. Intuitively, this approach allows Pastry to locate a node responsible for a certain key with logarithmic effort because in each routing step the matching prefix length is likely to be increased by one.

Since the prefix increases by one from routing table row to routing table row, there are also exponentially less candidates with which to fill a routing table entry as the row number increases. In Pastry, this leads to the effect (see [21, 23]) that from overlay routing step to overlay routing step the physical distance between nodes is likely to increase. Thus, the last routing step tends to dominate the overall physical routing path length of a key lookup.

Since the last overlay routing step is usually taken from the leaf set, with DynaMO this routing step is likely to be close. This is because leaf set entries are numerically closest to the current node, and thus they are also likely to be physically close to the current node due to DynaMO's ID assignment strategies. In other words, DynaMO promises to optimize the "last mile" of the overlay routing process.

4.3.1 Random Landmarking

Conventional landmarking, as introduced in [24, 25], suffers from the limitation that it assumes a set of fixed, stationary landmark nodes. All overlay nodes are expected to know the landmark nodes and to measure their respective distances to those landmarks. This, obviously, reintroduces the client-server concept into the bootstrap process. Especially in networks where nodes are expected to fail frequently, there are usually no sets of fixed nodes available, which renders this approach infeasible. Therefore, DynaMO introduces random landmarking (RLM) into the overlay construction process.

RLM utilizes the overlay lookup capabilities to locate overlay nodes responsible for a fixed set of landmark keys (overlay IDs). These nodes serve as temporary landmarks for a joining node. It is important to understand that the keys have to be chosen in a way that they divide the overlay ID space into equal portions. For example, in a network with an ID base of 16, an appropriate set of landmark keys would be: 000..00, 100..00, 200..00, ..., F00..00. The joining node then measures the distances to those temporary landmarks and assigns itself an ID based on its landmark ordering. The advantage of this approach is that "landmark nodes" can fail and others will simply step in as Pastry will automatically redirect future key lookups to those nodes now responsible for the landmark keys.

After having measured its landmark distances, the joining node adopts an ID prefix of a certain length from the landmark node closest[3] to itself. The ID remainder can be assigned randomly or can be based on an algorithm that further takes into account the physical neighborhood. The length of the ID prefix that the new node shares with its closest landmark node can be determined using the following formula:

$prefix\ length = [\log_b k]$

where b is the ID base and k the number of landmark keys. As can be seen, the number of landmark keys should preferably equal a power of b.

This approach has the following effects. First of all, it leads to physically close nodes forming regions with common ID prefixes, which means these nodes are also likely to be numerically close to each other in the overlay ID space, as can be seen in Figure 3. This, in turn, leads to the desired effect that a node's leaf set is likely to reference physically close nodes (bear in mind that the leaf set of a node contains the numerically closest nodes). Since the leaf set is normally utilized for the last routing step, that step is likely to travel a short physical distance. Note that there are still less and less candidate

---

[3] Conceivable metrics include hop count, RTT, etc.

nodes to choose from to fill a certain overlay routing table entry as the row number increases, but with DynaMO the likelihood of these candidates being physically close to the current node also increases from row to row.

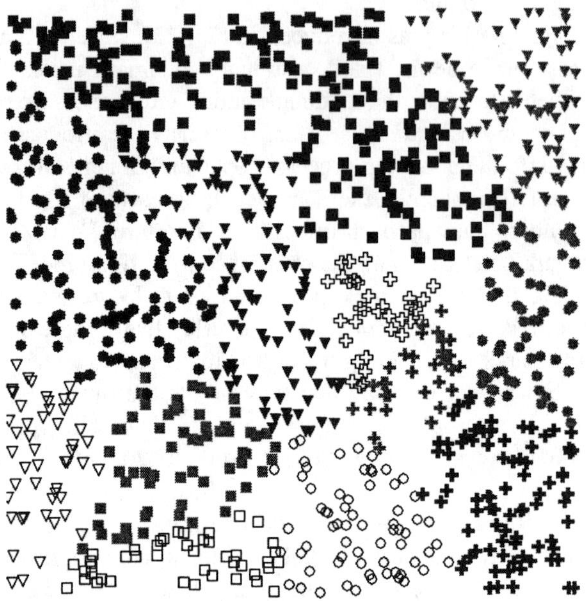

Figure 3: Spatial prefix distribution as generated by RLM. Equal symbols represent equal prefixes.

Special care has to be taken when a network is first created from scratch. To prevent temporary landmark nodes from being located too close to each other in the underlying network, the notion of a *landmark gravitation range* is introduced. If a new node discovers during its landmark measurement process that a temporary landmark node is responsible for a landmark key with which it shares no common prefix – i.e. that landmark must, therefore, be responsible for more than one landmark key – the new node should make itself a new landmark. However, it will only do so if its physical distance to *any* other landmark node exceeds a certain threshold, the landmark gravitation range. Again, various distance metrics are conceivable. The gravitation range is only a measure of reassurance that no physical landmark clusters form. It is therefore only significant during the initial network build-up because after all landmark keys are properly covered, this process ceases to have any importance.

To make the whole landmarking process more lightweight and efficient, a node obtains from its bootstrap node a list of the landmarks that the bootstrap node itself had used when it first joined the network. The idea here is that those "old" landmarks could still be valid. Thus, the new node is spared to initiate lookups for all landmark keys. The joining node now measures the distances to its inherited landmarks. When one of these landmarks receives the measure request from the joining node, it checks whether it is still responsible for the corresponding landmark key. If it is not, it will signal so in its measure response. Only in this case will the joining node reinitiate a landmark key lookup. Afterward, it will measure its distance to the proper landmark. This can reduce the overall bootstrap traffic significantly.

4.3.2 Experimental results

In order to examine DynaMO's behavior and performance, we simulated both a Pastry and a DynaMO network using the discrete event simulator Omnet++ [26]. Since we wanted to especially evaluate the overlay behavior and resilience in extremely dynamic networks, we chose to run Pastry and DynaMO overlays in ad-hoc scenarios employing an AODV [27] physical routing layer.

To put DynaMO's simulation results into perspective, we implemented a Pastry reference overlay in Omnet++ in strict conformance with the Pastry papers [21, 23]. Although these papers describe simulation results in detail, we implemented Pastry to have a reference basis which can be exposed to exactly the same networking conditions as our DynaMO implementation. Additionally, the result of our Pastry implementation can be compared to the paper results to give confidence in the correctness of DynaMO's implementation details.

We evaluated DynaMO in three main network settings: static networks, networks with degression and mobile networks. In a static network, the initial network topology remains unchanged over the entire course of a simulation run. In networks with degression, random nodes leave and join the network with a certain rate. In mobile networks, the set of participating nodes remains the same, but nodes are constantly changing their physical position. As our initial physical topology, we chose a plane where nodes are distributed randomly and with a certain density. For each network setting and scenario, we conducted multiple simulation runs.

In the following we will describe the mobile scenario only as this fits perfectly with our example of tourists walking through a city and looking for information provided by peer tourists. Mobile networks represent the biggest challenge when building topology-aware overlay networks because the underlying physical network changes constantly. In order to evaluate the performance of DynaMO in such networks, we conducted several simulations comparing Pastry and RLM. For RLM, we implemented an ID reassignment strategy to deal with mobility.

Pastry has no explicit mechanisms to deal with rapid topological changes in its underlying physical network. The only way to adapt its routing tables to reflect a modified physical underlay is to periodically run routing table maintenance tasks. These tasks are not run explicitly to detect mobility-induced changes, but instead are performed to compensate for any effects causing routing table deterioration.

The routing table maintenance as described in [23] performs the following task. In certain intervals, each node randomly selects an entry from each row of its routing table. It then asks each of those nodes for its corresponding routing table row. After that, it compares each entry in the row received with the corresponding entry in its own row by probing the distances to them. If need be, it replaces its own entry.

RLM uses a different strategy that deals explicitly with mobility-induced topology changes. Every node periodically re-measures its distance to the current landmark nodes. If its ID prefix is still congruent with the prefix of its closest landmark node, it will increase its re-measure interval by some factor. Otherwise, it will re-assign itself a new overlay ID based on the same strategy as used during its bootstrap and will rejoin the network with its new ID. Due to the extremely dynamic nature of mobile networks, a node uses the standard Pastry bootstrap optimization after it has rejoined the network under its new ID. This serves to propagate its new ID faster.

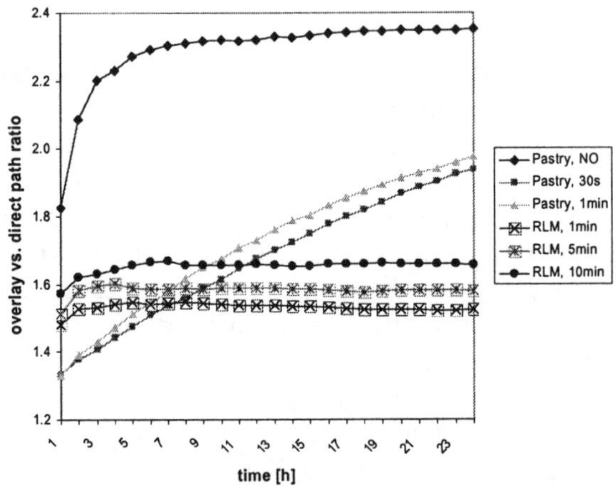

Figure 4: Overlay vs. direct path ratio change over time with Pastry and RLM.

Both Pastry and RLM were evaluated with the following network settings. Due to the added simulation complexity, a 2,000 node physical network served as the underlay. Each test run lasted 24 simulated hours, which we consider a typical stay in a city. During those 24h, 20,000 random lookups were issued. The nodes in the network moved according to the random waypoint mobility model [28] with a speed of 0.6m/s and a pause time of 30s.

For our Pastry experiments, we evaluated two different routing table maintenance intervals, as well as Pastry networks with no maintenance at all. Before mobility set in, all Pastry networks were artificially bootstrapped to start out with an optimal ratio of 1.31. As Figure 4 shows, if no routing table maintenance is performed, Pastry's ratio quickly deteriorates to a level of 2.35 that it would also roughly achieve in static networks without any optimization. Therefore, we considered next a maintenance interval of 1 minute so that each node runs the routing table maintenance task as explained above every minute. As can be seen, Pastry is unable to maintain a stable factor over time. Its ratio deteriorates nearly linearly until it reaches 1.97 after 24 hours. Clearly, this ratio does not peak here but would further rise. For this reason we conducted a second set of experiments, lowering the interval to 30s thereby increasing the maintenance effort markedly. The results indicate that even with this increased effort Pastry still fails to reach a stable ratio level after 24 hours with a ratio of 1.94 that is only negligibly lower than before.

With the same mobility parameters, we next evaluated RLM. Before mobility set in, all RLM networks were bootstrapped without optimization yielding an initial ratio of 1.37 slightly above Pastry's initial artificial ratio. Figure 4 shows that with a re-measure interval of 1 minute, RLM quickly reaches a stable ratio level of 1.54. If the re-measure interval is increased to 5 minutes, RLM achieves a stable ratio level of 1.58. Even if the nodes re-measure their landmark distances only every 10 minutes, RLM still maintains a stable ratio level of 1.65.

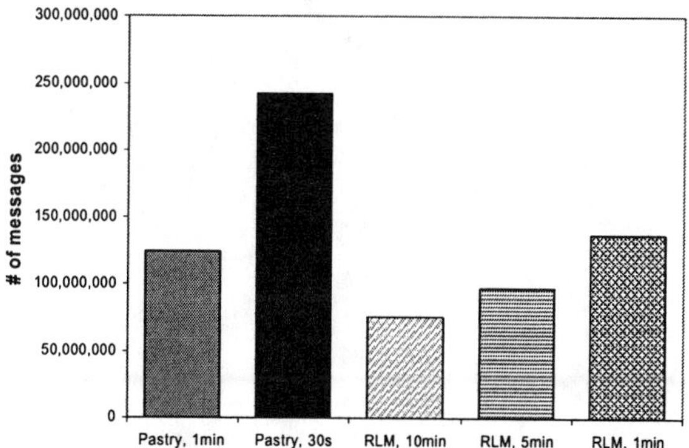

Figure 5: Number of overlay messages exchanged during an average 24h simulation.

Figure 5 shows the number of overlay messages exchanged during an average 24h simulation run. These messages include rejoin requests and forwards (only with RLM), distance measurements, and messages containing routing table state information. It shows that Pastry exchanges large volumes of overlay messages during the simulated 24h with both maintenance intervals (124 million and 241 million, respectively). Despite that, Pastry is not able to reach a stable ratio level after 24h as Figure 4 shows. RLM, on the other hand, achieves a stable ratio level (1.65) significantly better than Pastry's with significantly less messages exchanged (75 million) when the re-measure interval is 10 minutes. If one is willing to accept a message total slightly below or above Pastry's 1-minute interval total, RLM's ratio can be lowered even further. It has to be mentioned that in object storage overlays the rejoin of a node induces some additional overhead. Before a node assigns itself a new ID, it would have to pass on the references to the objects it is currently responsible for to its left and right neighbor in the ID space as they now become responsible for them. After rejoining the network with its new ID, the node would have to acquire the references to the objects it has now become responsible for from its new left and right ID space neighbor. Obviously, this overhead depends largely on the number of objects being stored on the network and the number of participating nodes.

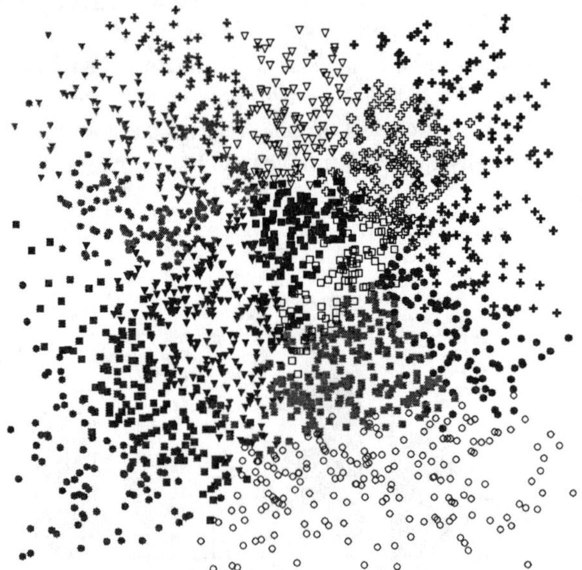

Figure 6: Spatial overlay ID distribution in an RLM-based mobile network with a 1-minute re-measure interval at the end of a simulation. Equal symbols represent equal prefixes

Figure 6 gives a visual interpretation of the spatial overlay ID distribution in RLM networks with a re-measure interval of 1 minute at the end of a 24h simulation. With the additional message overhead that a 1-minute interval generates, the spatial distribution remains nearly perfect. With the markedly smaller overhead of the much more coarse-grained 10-minute interval, the overlay ID clustering deteriorates slightly, but still remains acceptable.

4.4 Comparison with other approaches

A significant amount of work has been dedicated to the development of P2P overlay networks, but so far only few approaches explicitly focus on making overlay networks reflect the locality properties of the underlying physical networks. Furthermore, none of the related approaches consider node mobility.

One of the general concepts used to close the gap between physical and overlay node proximity is landmark clustering. Ratnasamy et al. [24] use landmark clustering in an approach to build a topology-aware CAN [19] overlay network. They require a fixed set of landmark nodes that all participating nodes have to know about. Prior to joining the overlay network, a joining node has to measure its distance (e.g., RTT, hop count, or any other appropriate metric) to each landmark. The node then orders the landmarks according to its distance measurements. Nodes with the same such landmark ordering fall into the same bin. The intuition behind this idea is that nodes with the same landmark ordering, i.e. nodes that have similar distances to all landmark nodes, are also quite likely to be close to each other topologically. Each bin is now mapped to a region in CAN's virtual coordinate space. After having *binned* itself, a joining node assumes a random point in the region associated with its bin. An immediate issue with landmark binning is that it can be rather coarse-grained depending on the number of landmarks used and their distribution. Furthermore, a fixed set of landmarks renders this approach unsuitable for dynamic networks, such as ad-hoc networks. The most significant downside of this approach, however, is that it can lead to an extremely uneven overlay ID distribution. This means that a small set of nodes could be responsible for a very large part of the ID space, essentially turning them into hot spots. Xu et al. [25] have verified this in their study.

Reference [25] presents a method to fine-tune landmark binning for the construction of overlay networks. They introduce maps containing information on nearby nodes in a specific region to allow nodes to join the overlay network with a more accurate reflection of its own position in the physical network. A map is stored as global soft state among the nodes of a region. This approach, however, comes with a significant overhead. Potentially, there could be a *very* large number of regions (e.g., in Pastry such a region is considered to be a set of nodes sharing a certain ID prefix), all of which have to maintain their map. Moreover, to achieve a finer granularity, additional inner-bin measurements are required.

Waldvogel and Rinaldi [29] propose an overlay network (Mithos) that focuses on reducing routing table sizes. Mithos also tries to establish overlay locality from physical network proximity. A new node is assigned an overlay ID based on the IDs of its (physical) neighbors. They employ virtual springs to make the ID fit into the neighborhood range. In order to avoid local minima, substantial probing has to be undertaken. Unfortunately, only very small overlay networks (200 – 1000 nodes) are used for simulations and the impact of network degression is not considered.

As a DHT, Pastry [21] uses certain heuristics to exploit physical network proximity in its overlay routing tables. In a thorough analysis [23], Castro et al. examine the impact of various network parameters and node degression on Pastry's locality properties. Unlike the other approaches presented, Pastry does not construct its overlay structure from the underlying physical network topology. Instead, Pastry distributes its node evenly in the overlay ID space regardless of the actual physical topology. One way in which Pastry tries to exploit physical proximity is that a new node should bootstrap itself using a node nearby. During the join process, it then tries to choose among the candidate nodes for particular routing table entry a node that is "close" to itself. During its lifetime, a node periodically performs routing table maintenance and improvement by asking other nodes for "better" routing table entries. Obviously, those are mere heuristics and, therefore, Pastry does not guarantee optimal routing table states.

5. CONCLUSION

This paper focused on the delivery of information to mobile users and on the exchange of information among them. We concentrated on the main general concepts of such applications as well as on the common infrastructures to supports them. Note that many general concerns of mobile applications were not discussed here even though they are of prime importance. They include privacy, security, service access, possible chaining of services, and pricing.

We took as a reference an application from information delivery in tourism as it encompasses a rich set of functionalities, among them: classical information delivery, scalability issues, refinements of delivered information, adaptation to the profile of the end users (personalization), and information exchange with peers. The latter is a relatively new concept which has, to our knowledge, not yet been implemented in such applications. A simple functionality of a general tourism information provider is to answer the query of an end user, e.g., "where can I find a medium-budget Italian restaurant around here?" The service that handles this query is a location-based service (LBS

[30]) that knows the position of the user, links it to the location of relevant restaurants (combination of maps and yellow pages), and possibly displays to the user the route to get there. This is a typical *pull* mechanism, where the user pulls information from the system. A dual mechanism is that of the system *pushing* information to the user when it judges it appropriate, for instance to point to a mobile tourist a place that can be of interest to him or her because it corresponds to his/her profile and is nearby.

In mobile applications, delivering information is related to event notification systems (ENS), where an ENS takes into account an event that occurred and executes an action associated to that event. For instance, in our context, if the user has seen all museums of a city except one he/she may be interested in seeing the only one that was not seen yet. This is part of the application profile. In this case, an event is the fact that the user has seen yet another museum. A more general, common event is a change of location from the user, and which is detected periodically (with an epoch defined in a system). In our approach, we extended the classical notion of events to handle such cases.

We first presented an exemplary application, i.e., an enhanced tourist application with two mobile users asking for information and receiving information independently from each other and who communicate with other users to *exchange* information. We then discussed the classical centralized approach, server client like where client may pose LBS queries to the server and where the server delivers relevant information. We illustrated it with the TIP system (Tourism Information Provider) developed at FU Berlin. Another paradigm in this context is that of *geocasting* where a source of information sends to all users around some information which is filtered locally, i.e., on the mobile device. In our scenario, this occurs when the user passes by the ice cream place. The information is sent from an access point to all the mobile devices in a radius of about typically 40 m. The local system checks whether the information is of any relevance to the user. This is matched against his/her profile (specified of inferred). Mobile devices being more and more powerful, in terms of autonomy and computing power, it is our belief that we will soon move away from these two classical paradigms and tend to a mix of the two solutions, with many functionalities implemented locally.

We then focused on the information exchange among user, using peer-2-peer networks. This highly-decentralized infrastructure allows groups of users – being defined geographically or around a particular interest – to exchange information quickly. Of course this paradigm is based on incentives to share information (i.e., altruistic models). The major challenge of our context is that we need to consider *mobile* P2P applications, hence dynamic configurations, which leads to performance concerns. We explained the major issues as well as the solution currently under development at FU Berlin (DynaMO). Compared to Pastry, the basis for our approaches, we were able to achieve comparable ratios with significantly less communication effort. This is particularly important in mobile environments with battery driven devices. Up to 5 to 7 times more messages have to be exchanged using Pastry compared to RLM achieving the same ratio. On the other hand, when adding the same optimization effort as with Pastry, our approaches were able to achieve a routing ratio decidedly lower than even the artificial Pastry optimum. Furthermore, we were able to preserve an even ID distribution employing RLM. Our second goal was to design our approaches in a way that they can maintain their key properties even in the presence of high network dynamics, especially node mobility. We showed that DynaMO maintains a significantly lower overlay vs. direct path ratio than Pastry does in mobile networks while exchanging significantly less messages. We believe that with the development of peer-to-peer networks, such approaches will be common features used by mobile users in the near future.

6. ACKNOWLEDGMENTS

The authors want to thank Annika Hinze (University of Waikato, Hamilton, New Zealand), Rolf Winter and Thomas Zahn (both of Freie Universität Berlin, Berlin, Germany) for their valuable contributions to TIP and DynaMO.

7. REFERENCES

1. Hinze, A. and Voisard, A. Location- and Time-Based Information Delivery in Tourism, in *Advances in Spatial and Temporal Databases*, in Proc. of the Intl. Conference on Spatio-temporal Databases, LNCS No. 2750, Springer Verlag, Berlin, 2003.
2. Volz, S. and Klinec, D., Nexus: The Development of a Platform for Location Aware Application, in P*roc. of the Third Turkish-German Joint Geodetic Days Towards a Digital Age*, 1999.
3. Cheverst, K., Mitchell, K., and Davies, N., The Role of Adaptive Hypermedia in a Context-Aware Tourist GUIDE, *Communications of the ACM*, 45(5):47–51, 2002.
4. Poslad, S., Laamanen, H., Malaka, R., Nick, A., Buckle, P., and Zipf, A., Crumpet: Creation of User-Friendly Mobile Services Personalized for Tourism, *Proc. of 3G 2001 - Second International Conference on 3G Mobile Communication Technologies*, 2001.
5. Brinkhoff, T. and Weitkämper, J., Continuous Queries Within an Architecture for Querying XML-Represented Moving Objects, P*roc. of the 7th Intl. Symposium on Spatial and Temporal Databases* (SSTD), LNCS No. 2121, Springer Verlag, Berlin, 2001.

6. Pashtan, A., Blattler, R., Heusser, A., and Scheuermann, P., CATIS: A Context-Aware Tourist Information System, P*roc. 4th Int. Workshop of Mobile Computing*, 2003.
7. Oppermann, R., Specht, M., and Jaceniak, I., Hippie: A Nomadic Information System, P*roc. of the First International Symposium Handheld and Ubiquitous Computing* (HUC), 1999.
8. His, S., The Electronic Guidebook: A Study of User Experiences Using Mobile Web Content in a Museum Setting, *Proc. of the IEEE Intl. Workshop on Wireless and Mobile Technologies in Education* (WMTE), 2002.
9. Fleck, M., Frid, M., Kindberg, T., O'Brien-Strain, E., Rajani, R., and Spasojevic, M., From Informing to Remembering: Ubiquitous Systems in Interactive Museums, *Pervasive Computing*, 1(2):13–21, 2002.
10. Sistla, A. P., Wolfson, O., Chamberlain, S., and Dao, S. Modeling and Querying Moving Objects. ICDE, pp. 422–432, 1997.
11. Vazirgiannis, M. and Wolfson, O., A spatiotemporal model and language for moving objects on road networks, proc. of the 7th Intl. Symposium on Spatial and Temporal Databases (SSTD), LNCS no. 2121, Springer Verlag, 2001.
12. Su, J., Xu, H., and Ibarra, O. H., Moving objects: Logical Relationships and Queries, *Proc. of the 7th Intl. Symposium on Spatial and Temporal Databases* (SSTD), LNCS No. 2121, Springer Verlag Heidelberg/ Berlin/ New York, 2001.
13. Forlizzi, L., Güting, R. H., Nardelli, E., and Schneider, M., A Data Model and Data Structures for Moving Objects Databases, *Proc. of the ACM SIGMOD Conf. on Management of Data*, 2000.
14. Güting, R. H., Böhlen, M. H., Erwig, M., Jensen, C. S., Lorentzos, N. A., Schneider, M., and Vazirgiannis, M., A Foundation for Representing and Quering Moving Objects, *Transactions of Database Systems* (ACM TODS), 25(1):1–42, 2000.
15. Hinze, A. and Voisard, A., A Parameterized Algebra for Event Notification Services, *Proc. of the 9th Intl. Symposium on Temporal Representation and Reasoning* (TIME), 2002.
16. Schiller, J. *Mobile communications.* 2nd Edition. Addison Wesley, London 2003.
17. The Gnutella Protocol Specification v0.4., www9.limewire.com/developer/ gnutella_protocol_0.4.pdf.
18. Ritter, J., Why Gnutella Can't Scale. No, Really., http://www.darkridge.com/~jpr5/doc/ gnutella.html.
19. Ratnasamy, S., Francis, P., Handley, M., Karp, R., and Shenker, S., A Scalable Content-Addressable Network, *Proc. of ACM SIGCOMM*, August 2001.

20. Stoica, I., Morris, R., Karger, D., Kaashoek, M. F., and Balakrishnan, H., Chord: A scalable peer-to-peer lookup service for internet applications, proc. of ACM SIGCOMM, August 2001.
21. Rowstron, A. and Druschel, P., Pastry: Scalable, Decentralized Object Location and Routing for Large-Scale Peer-to-Peer Systems, in *Proc. Intl. Conference on Distributed Systems Platforms* (Middleware 2001), 2001.
22. Zhao, B. Y., Kubiatowicz, J. D., and Joseph, A. D., Tapestry: An Infrastructure for Fault-Resilient Wide-Area Location and Routing, Technical Report UCB//CSD-01-1141, U.C. Berkeley, CA, April 2001.
23. Castro, M., Druschel, P., Hu, Y. C., and Rowstron, A., Exploiting Network Proximity in Peer-to-Peer Overlay Networks, International Workshop on Future Directions in Distributed Computing (FuDiCo), 2002.
24. Ratnasamy, S., Handley, M., Karp, R., and Shenker, S., Topologically-Aware Overlay Construction and Server Selection, *IEEE Infocom* '2002.
25. Xu, Z., Tang, C., and Zhang, Z., Building Topology-Aware Overlays using Global Soft-State, *ICDSC*, 2003.
26. Omnet++, Discrete Event Simulation System, http://www.omnetpp.org.
27. Ad hoc On-Demand Distance Vector Routing, http://www.ietf.org.
28. Johnson, D. B. and Maltz, D. A., Dynamic Source Routing in Ad hoc Wireless Networks, In *Mobile Computing*, Imielinski and Korth (Eds.), Kluwer Academic Publishers, 1996, Vol. 353.
29. Waldvogel, M. and Rinaldi, R., Efficient Topology-Aware Overlay Network, HotNets 2002, *SIGCOMM/CCR* 2003.
30. Schiller, J. and Voisard, A. (Eds.), *Location-Based Services*, Morgan Kaufmann, San Francisco, to appear in Spring 2004.

Image Processing and Sensor Networks

Feature-Based Georegistration of Aerial Images

Yaser Sheikh[1], Sohaib Khan[2] and Mubarak Shah[1]

[1] Computer Vision Lab,
School of Computer Science,
University of Central Florida,
Orlando, FL 32816-2362,
USA

[2] Department of Computer Science and Computer Engineering,
Lahore University of Management Sciences,
Lahore, Pakistan

1 Introduction

Georegistration is the alignment of an observed image with a geodetically calibrated reference image. Such alignment allows each observed image pixel to inherit the coordinates and elevation of the reference pixel it is aligned to. Accurate georegistration of video has far-reaching implications for the future of automation. An agent (such as a robot or a UAV), equipped with the ability to precisely assign geodetic coordinates to objects or artifacts within its field of view, can be an indispensable tool in applications as diverse as planetary exploration and automated vacuum cleaners. In this chapter, we present an algorithm for the automated registration of aerial video frames to a wide area reference image. The data typically available in this application are the reference imagery, the video imagery and the telemetry information.

The reference imagery is usually a wide area, high-resolution ortho-image. Each pixel in the reference image has a longitude, latitude and elevation associated with it (in the form of a Digital Elevation Map - DEM). Since the reference image is usually dated by the time it is used for georegistration, it contains significant dissimilarities with respect to the aerial video data. The aerial video data is captured from a camera mounted on an aircraft. The orientation and position of the camera are recorded, per-frame, in the telemetry information. Since each frame has this telemetry information associated with it, georegistration would seem to be a trivial task of projecting the image onto the reference image coordinates. Unfortunately, mechanical noise causes fluctuations in the telemetry measurements, which in turn causes significant projection errors, sometimes up to hundreds of pixels. Thus while the telemetry information provides *coarse* alignment of the video frame, georegistration techniques are required to obtain accurate pixel-wise calibration of each aerial image pixel. In this chapter, we use the telemetry information to orthorectify the aerial images, to bring both imageries into a common projection space, and then apply our registration technique to achieve accurate alignment. The challenge in georegistration lies in the stark differences between the video and reference data. While the difference of projection view is accounted for by orthorectification, four types of data distortions are still encountered: (1) Sensor noise in the form of erroneous telemetry data, (2) Lighting and atmospheric changes, (3) Blurring, (4) Object changes in the form of forest growths or new construction. It should also be noted that remotely sensed terrain imagery has the property of being highly self-correlated both as image data and elevation data. This includes first order correlations (locally similar luminance or elevation values in buildings), second order correlations (edge continuations in roads, forest edges, and ridges), as well as higher order correlations (homogeneous textures in forests and homogenous elevations in plateaus). Therefore, while developing georegistration algorithms the important criterion is the robust handling of outliers caused by this high degree of self-correlation.

1.1 Previous Work

Currently several systems that use geolocation have already been deployed and tested, such as Terrain Contour Matching (TERCOM) [10], SITAN, Inertial Navigation / Guidance Systems (INS/IGS), Global Positioning Systems (GPS) and most recently Digital Scene-Matching and Area Correlation (DSMAC). Due to the limited success of these systems and better understanding of their shortcomings, georegistration has recently received a flurry of research attention. Image-based geolocation (usually in the form of georegistration) has two principal properties that make them of interest: (1) Image capture and alignment is essentially a passive application that does not rely on interceptable emissions (like GPS systems) and (2) Georegistration allows independent per-frame geolocation thus avoiding cumulative errors. Image based techniques can be broadly classified into two approaches: Intensity-based approaches and elevation-based approaches.

The overriding drawback of elevation-based approaches is that they rely on the accuracy of recovered elevation from two frames, which has been found to be difficult and unreliable. Elevation based algorithms achieve alignment by matching the reference elevation map with an elevation map recovered from video data. Rodrequez and Aggarwal in [24] perform pixel-wise stereo analysis of successive frames to yield a recovered elevation map or REM. A common representation ('cliff maps'), are used and local extrema in curvature are detected to define critical points. To achieve correspondence, each critical point in the REM is then compared to each critical point in the DEM. From each match, a transformation between REM and DEM contours can be recovered. After transforming the REM cliff map by this transformation, alignment verification is performed by finding the fraction of transformed REM critical points that lie near DEM critical points of similar orientation. While this algorithm is efficient, it runs into similar problems as TERCOM i.e. it is likely to fail in plateaus/ridges and depends highly on the accurate reconstruction of the REM. Finally, no solution was proposed for computing elevation from video data. More recently in ([25]), a relative position estimation algorithm is applied between two successive video frames, and their transformation is recovered using point-matching in stereo. As the error may accumulate while calculating relative position between one frame and the last, an absolute position estimation algorithm is proposed using image based registration in unison with elevation based registration. The image based alignment uses Hausdorff Distance Matching between edges detected in the images. The elevation based approach estimates the absolute position, by calculating the variance of displacements. These algorithms, while having been shown to be highly efficient, restrict degrees of alignment to only two (translation along x and y), and furthermore do not address the conventional issues associated with elevation recovery from stereo.

Image-based registration, on the other hand, is a well-studied area. A somewhat outdated review of work in this field is available in [4]. Conventional alignment techniques are liable to fail because of the inherent differences between the two imageries we are interested in, since many corresponding pixels are often dissimilar. Mutual Information is another popular similarity measure, [30], and while it provides high levels of robustness it also allows many false positives when matching over a search area of the nature encountered in georegistration. Furthermore, formulating an efficient search strategy is difficult. Work has also been done in developing image-based techniques for the alignment of two sets of reference imageries [32], as well as the registration of two successive video images ([3], [27]). Specific to georegistration, several intensity based approaches to georegistration intensity have been proposed. In [6], Cannata *et al* use the telemetry information to bring a video frame into an orthographic projection view, by associating each pixel with an elevation value from the DEM. As the telemetry information is noisy the association of elevation is erroneous as well. However, for aerial imagery that is taken from high altitude aircrafts the rate of change in elevation may be assumed low enough for the elevation error to be small. By orthorectifying the aerial video frame, the process of alignment is simplified to a strict 2D registration problem. Correspondence is computed by taking 32×32 pixel patches uniformly over the aerial image and correlating them with a larger search patch in the Reference Image, using Normalized Cross Correlation. As the correlation surface is expected to have a significant number of outliers, four of the strongest peaks in each correlation surface are selected and consistency measured to find the best subset of peaks that may be expressed by a four parameter affine transform. Finally, the sensor parameters are updated using a conjugate gradient method, or by a Kalman Filter to stress temporal continuity. An alternate approach is presented by Kumar *et al* in [18] and by Wildes *et al* in [31] following up on that work, where instead of ortho-rectifying the Aerial Video Frame, a perspective projection of the associated area of the Reference Image is performed. In [18], two further data rectification steps are performed. Video frame-to-frame alignment is used to create a mosaic providing greater context for alignment than a single image. For data rectification, a Laplacian filter at multiple scales is then applied to both the video mosaic and reference image. To achieve correspondence, coarse alignment is followed by fine alignment. For coarse alignment feature points are defined as the locations where the response in both scale and space is maximum. Normalized correlation is used as a match measure between salient points and the associated reference patch. One feature point is picked as a reference, and the correlation surfaces for each feature point are then translated to be centered at the reference feature point. In effect, all the correlation surfaces are superimposed, and for each location on the resulting superimposed surface, the top k values (where k is a constant dependent on number of feature points) are multiplied together to establish a consensus surface. The highest resulting point on the correlation surface is then taken to be the true displacement. To achieve fine alignment, a 'direct' method of alignment is employed, minimizing the SSD of user selected areas in the video and reference (filtered) image. The plane-parallax model is employed, expressing the transformation between images in terms of 11 parameters, and optimization is achieved iteratively using the Levenberg-Marquardt technique.

In the subsequent work, [31], the filter is modified to use the Laplacian of Gaussian filter as well as it's Hilbert Transform, in four directions to yield four oriented energy images for each aerial video frame, and for each perspectively projected reference image. Instead of considering video mosaics for alignment, the authors use a mosaic of 3 'key-frames' from the data stream, each with at least 50 percent overlap. For correspondence, once again a local-global alignment process is used. For local alignment, individual frames are aligned using a three-stage Gaussian pyramid. Tiles centered around feature points from the aerial video frame are correlated with associated patches from the projected reference image. From the correlation surface the dominant peak is expressed by its covariance structure. As outliers are common, RANSAC is applied for each frame on the covariance structures to detect matches consistent to the alignment model. Global alignment is then performed using both the frame to frame correspondence as well as the frame-to-reference correspondence, in three stages of progressive alignment models. A purely translational model is used at the coarsest level, an affine model is then used at the intermediate level, and finally a projective model is used for alignment. To estimate these parameters an error function relating the Euclidean distances of the frame-to-frame and frame-to-reference correspondences is minimized using the Levenberg-Marquardt optimization.

1.2 Our Work

The focus of this paper is the registration of single frames, which can be extended easily to include multiple frames. Elevation based approaches were avoided in favor of image-based methods due to the unreliability of elevation recovery algorithms, especially in the self-correlated terrains typically encountered. It was observed that the georegistration task is a composite problem, most dependent on a robust correspondence module which in turn requires the effective handling of outliers. While previous works have instituted some outlier handling mechanisms, they typically involve disregarding some correlation information. As outliers are such a common phenomenon, the retention of as much correlation information as possible is required, while maintaining efficiency for real-time implementation. The contribution of this work is the presentation of a feature-based alignment method that searches over the entire set of correlation surface on the basis of a relevant transformation model. As the georegistration is a composite system, greater consistency in correspondence directly translates into greater accuracy in alignment. The algorithm described has three major improvements over previous works: Firstly, it selects patches on the basis of their intensity values rather than through uniform grid distributions, thus avoiding outliers in homogenous areas. Secondly, relative strengths of correlation surfaces are considered, so that the degree of correlation is a pivotal factor in the selection of consistent alignment. Finally, complete correlation information retention is achieved, avoiding the loss of data by selection of dominant peaks. By searching over the entire set of correlation surfaces it becomes possible not only to handle outliers, but also to handle the 'aperture effects' effectively. The results demonstrate that the proposed algorithm is capable of handling difficult georegistration problems and is robust to outliers as well.

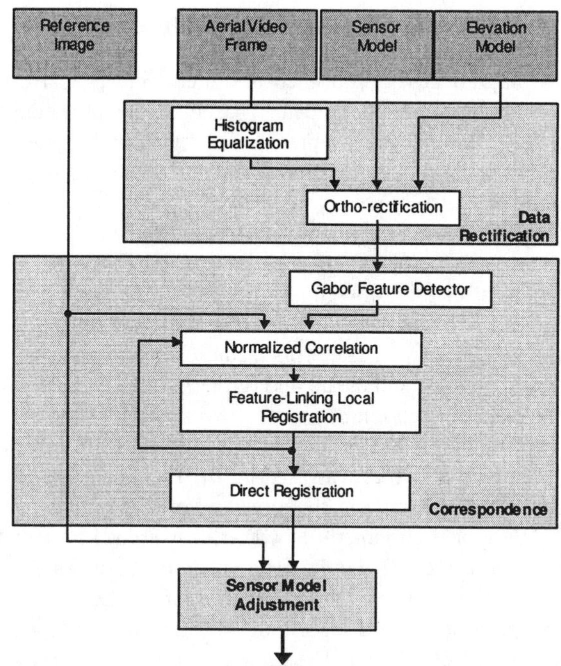

Fig. 1. A diagrammatical representation of the workflow of the proposed alignment algorithm. The four darker gray boxes (Reference Image, Aerial Video Frame, Sensor Model, and Elevation Model) represent the four inputs to the system. The three processes of Data Rectification, Correspondence and Model Update are shown as well.

The structure of the complete system is shown in Figure 1. In the first module Projection View rectification is performed by the orthographic projection of the Aerial Video Image. This approach is chosen over the perspective projection of the reference image to simplify the alignment model, especially since the camera attitude is approximately nadir, and the rate of elevation change is fairly low. Once both images are in a common projection view, feature-based registration is performed by linking correlation surfaces for salient features on the basis of a transformation model followed by direct registration within a single pyramid. Finally, the sensor model parameters are updated on the basis of the alignment achieved, and the next frame is then processed.

The remainder of this chapter is organized as follows. In Section 2 the proposed algorithm for feature-based georegistration is introduced, along with an explanation of feature selection and feature alignment methods. Section 3 discusses the sensor parameter update methods. Results are shown in Section 4 followed by conclusions in Section 5.

2 Image Registration

In this paper, alignment is approached in a hierarchical (coarse-to-fine) manner, using a four level Gaussian pyramid. Feature-based alignment is performed at coarser levels of resolution, followed by direct pixel-based registration at the finest level of resolution. The initial feature-matching is important due to the lack of any distinct global correlation (regular or statistical) between the two imageries. As a result,"direct" alignment techniques, i.e. techniques globally minimizing intensity difference using the brightness constancy constraint, fail on such images since global constraints are often violated in the context of this problem. However, within small patches that contain corresponding image features, statistical correlation is significantly higher. The selection of a similarity measure was normalized cross correlation as it is invariant to localized changes in contrast and mean, and furthermore in a small window it linearly approximates the statistical correlation of the two signals. Feature matching may be approached in two manners. The first approach is to select uniformly distributed pixels (or patches) as matching points as was used in [6]. The advantage of this approach is that pixels, which act as constraints, are spread all over the image, and can therefore be used to calculate global alignment. However, it is argued here that uniformly selected pixels may not necessarily be the most suited to registration, as their selection is not based on actual properties of the pixels intensities themselves (other than their location). For the purposes of this algorithm, selection of points was based on their response to a feature selector. The proposition is that these high response features are more likely to be matched correctly and would therefore lend robustness to the entire process. Furthermore, it is desirable in alignment to have no correspondences at all in a region, rather than have inaccurate ones for it. Because large areas of the image can potentially be textured, blind uniform selection often finds more false matches than genuine ones. To ensure that there is adequate distribution of independent constraints we pick adequately distributed local maximas in the feature space. Figure 2 illustrates the difference between using uniformly distributed points (a) and feature points (b). All selected features lie at buildings, road edges, intersections, points of inflexion etc.

2.1 Feature Selection

As a general rule, features should be independent, computationally inexpensive, robust, insensitive to minor distortions and variations, and rotational invariant. Additionally, one important consideration must be made in particular for the selection of features for remotely sensed land imageries. It has already been mentioned that terrain imagery is highly self-correlated, due to continuous artifacts like roads, forests, water bodies etc. The selection of the basic features should be therefore related to the compactness of signal representation. This means a representation is sought where features are selected that are not locally self-correlated, and it is intuitive that in normalized correlation between the Aerial and Reference Image such features would also have a greater probability of achieving a correct match. In this paper, Gabor Filters are used since they provide such a representation for real signals [9].

Gabor filters are directional weighted sinusoidals convoluted by a Gaussian window, centered at the origins (in two dimensions) with the Dirac function. They are defined as:

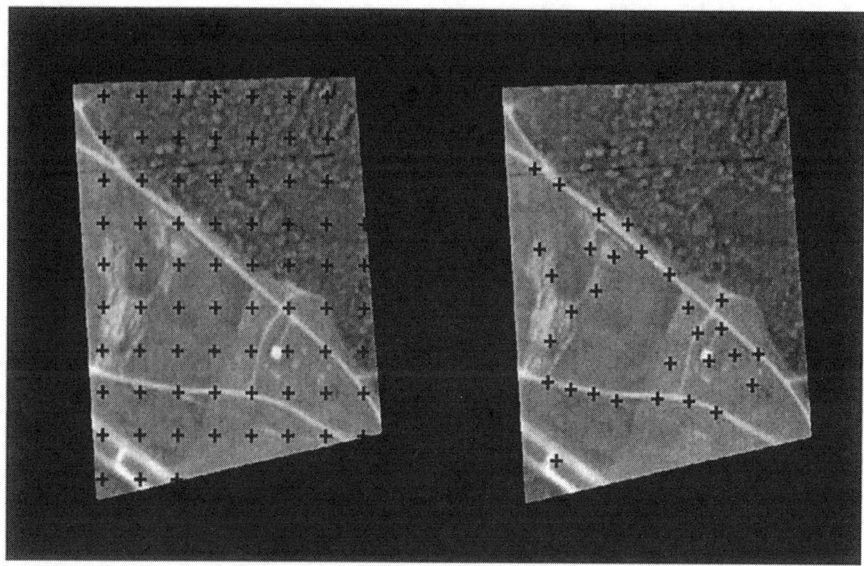

Fig. 2. Perspective projection of the reference image. (a) The aerial video frame displays what the camera *actually* captured during the mission. (b) Orthographic footprint of the aerial video frame on the reference imagery (c) The perspective projection of reference imagery displays what the camera *should* have captured according to the telemetry.

$$G(x, y, \theta, f) = e^{i(f_x x + f_y y)} e^{-(f_x^2 + f_y^2)(x^2+y^2)/2\sigma^2} \tag{1}$$

where x and y are pixel coordinates, $i = \sqrt{-1}$, f is the central frequency, q is the filter orientation, $f_x = f\cos\theta$, $f_y = f\sin\theta$, and s is the variance of the Gaussian window. Fig. 3 shows the four orientations of Gabor filter that were used for feature detection on the Aerial Video Frame. The directional filter responses were multiplied to provide a consensus feature surface for selection. To ensure that the features weren't clustered to provide misleading localized constraints, distributed local maximas were picked from the final feature surface. The particular feature points selected are shown in Figure 4. It is worth noting that even in the presence of significant cloud cover, and for occlusion by vehicle parts, in which the uniform selection of feature points would be liable to fail, the algorithm manages to recover points of interest correctly.

2.2 Robust Local Alignment

It is often over-looked that a composite system like georegistration cannot be any better than the weakest of its components. Coherency in correspondence is often the point of failure for many georegistration approaches. To address this issue a new transformation model based correspondence approach is presented in the orthographic projection view, however this approach may easily be extended to more general projection views and transformation models. Transformations in the orthographic viewing space are most closely modelled by affine transforms, as orthography accurately satisfies the

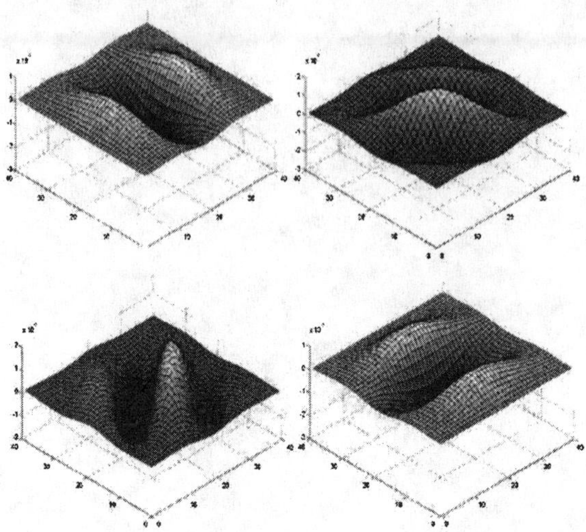

Fig. 3. Gabor filters are directional weighted sinusoidals convoluted by a Gaussian window. Four orientations of the Gabor filter are displayed.

weak-perspective assumption of the affine-model. Furthermore, the weak perspective model may also compensate for some minor errors introduced due to inaccurate elevation mapping. In general, transformation models may be expressed as

$$\mathbf{U}(\mathbf{x}) = T \cdot \mathbf{X}(\mathbf{x}) \tag{2}$$

where $\mathbf{U}$ is the motion vector, $\mathbf{X}$ is the pixel coordinate based matrix, and T is a matrix determined by the transformation model. For the affine case particularly, the transformation model has six parameters:

$$u(x, y) = a_1 x + a_2 y + a_3 \tag{3}$$
$$v(x, y) = a_4 x + a_5 y + a_6 \tag{4}$$

where u and v are the motion vectors in the horizontal and vertical directions. The six parameters of affine transformation are represented by the vector $\mathbf{a}$,

$$\mathbf{a} = [a_1 \;\; a_2 \;\; a_3 \;\; a_4 \;\; a_5 \;\; a_6]$$

If a planar assumption (the relationship between the two images is planar) is made to simplify calculation, the choice of an orthographic viewing space proves to be superior to the perspective viewing space. All the possible transformations in the orthographic space can be accurately modelled using six parameters of the affine model, and it is easier to compute these parameters robustly compared to a possible twelve-parameter model of planar-perspective transformation (especially since the displacement can be

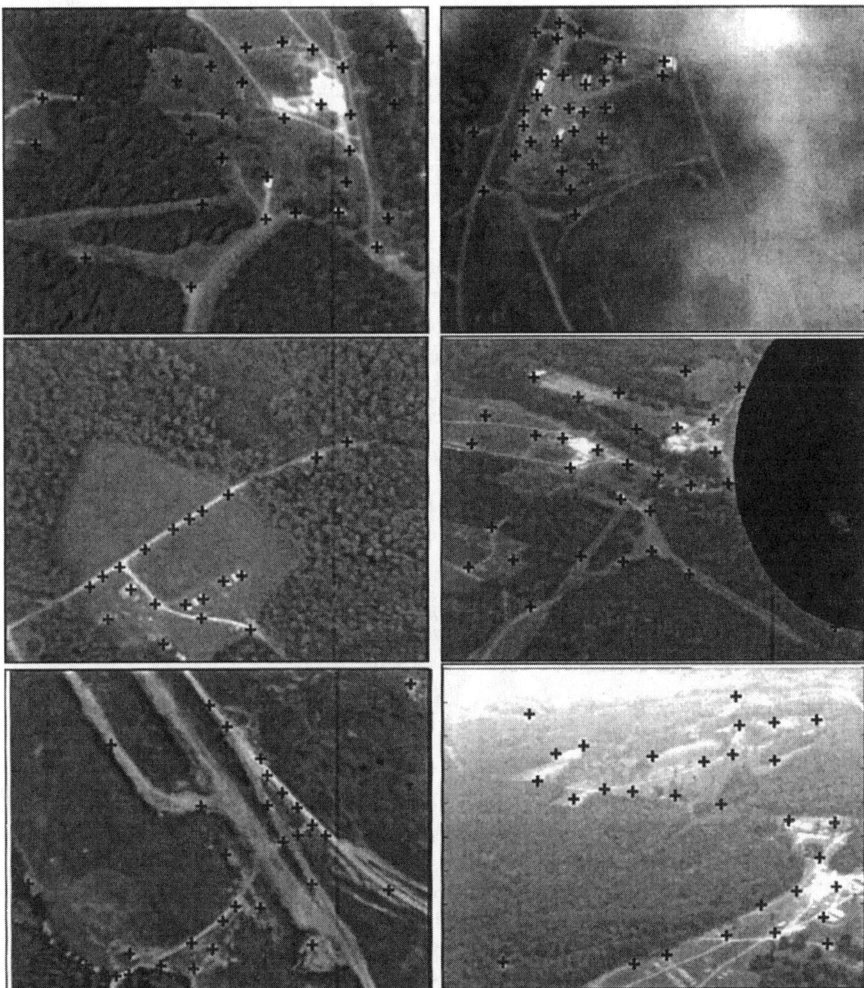

Fig. 4. Examples of features selected in challenging situations. Feature points are indicated by the black '+'s. Points detected as areas of high interest in the Gabor Response Image. Features are used in the correspondence module to ensure that self-correlated areas of the images do not contribute outliers. Despite cloud cover, occlusion by aircraft wheel, and blurring, salient points are selected. These conditions would otherwise cause large outliers and consequently leads to alignment failure.

quite significant). Furthermore, making a planarity assumption for a perspective projection view undermines the benefits of reference projection accuracy. Also, since the displacement between images can be up to hundreds of pixels, the fewer the parameters to estimate the greater the robustness of the algorithm. The affine transformation is estimated in a hierarchical manner, in a four-level Gaussian pyramid. At the lower resolution levels, the feature-matching algorithm compensates for the large displacements, while a direct method of alignment is used at the finest resolution levels so that information is not lost.

Feature Based Alignment

The Gabor Feature Detector returns n feature points (typically set to find between ten and twenty), to be used in the feature-based registration process. A patch around each feature pixel of the Aerial Video Frame is then correlated with a larger search window from the Cropped Reference Image to yield n correlation surfaces. For T_i, the patch around a feature point, the correlation surface is defined by normalized cross-correlation. For any pair of images $I_2(\mathbf{x})$ and $I_1(\mathbf{x})$, the correlation coefficient r_{ij} between two patches centered at location (x_i, y_j) is defined as

$$r(i,j) = \frac{\sum_{w_x} \sum_{w_y} (\phi_2)(\phi_1)}{\sqrt{\sum_{w_x} \sum_{w_y} (\phi_2)^2 \sum_{w_x} \sum_{w_y} (\phi_1)^2}} \tag{5}$$

where

$$\phi_1 = I_1(\mathbf{x} + [w_x \ w_y]^T) - \mu_1 \tag{6}$$

$$\phi_2 = I_2(\mathbf{x} + [w_x \ w_y]^T) - \mu_2 \tag{7}$$

and w_x and w_y are the dimensions of the local patch around (x_i, y_j), and μ_1 and μ_2 are the patch sample means.

To formally express the subsequent process of alignment, two coordinate systems are defined for the correlation surface. Each element on a correlation surface has a relative coordinate position (u, v), and an absolute coordinate position $(x_f - u, y_f - v)$, where (x_f, y_f) is the image coordinate of the feature point associated with each surface. The relative coordinate (u, v) of a correlation element is the position relative to the feature point around which the correlation surface was centered and the absolute position of the correlation surface is the position of each element on the image coordinate axes. Each correlation element $\eta_i(u, v)$ can be considered as a magnitude of similarity for the transformation vector from the feature point coordinate (x_f, y_f), to the absolute position of the correlation element $(x_f - u, y_f - v)$. Figure 5 (b) shows the absolute coordinate system and Figure 5 (c) shows the relative positions of each correlation element. Peaks in the correlation surfaces denote points at which there is a high probability of a match, but due to the nature of the Aerial Video Frame and the Reference Image discussed earlier each surface may include multiple peaks or ridges. Now, had the set of possible alignment transformations been only translational, the ideal consensus transformation could have been calculated by observing the peak in the element-wise sum (or product) of the n correlation surfaces. This 'sum-surface' $\eta(u, v)$ is defined over the relative coordinate system as,

$$\eta(u, v) = \sum_{i=1}^{n} \eta_1(u, v) \tag{8}$$

On this 'sum-surface', by picking the translation vector in the relative coordinate system, from the center to the maximum peak the alignment transformation can be recovered. It can also be observed that since translation is a position invariant transform (i.e. translation has the same displacement effect on pixels irrespective of absolute location) the individual correlation surfaces can be treated independent of their horizontal and vertical coordinates. Therefore the search strategy for finding the optimal translational transformation across all the n correlations is simply finding the pixel coordinates (u_{peak}, v_{peak}) of the highest peak on the Sum-Surface. Put another way, a translational vector is selected such that if it were applied simultaneously to all the correlation surfaces, the sum of values of the center position would be maximized. When the vector (u_{peak}, v_{peak}) is applied to the correlation surface in the relative coordinate system, it can be observed that $\eta(0,0)$ would be maximized for

$$\eta(u,v) = \sum_{i=1}^{n} \eta_i(u',v') \tag{9}$$

where

$$u' = u - u_{peak} \tag{10}$$

$$v' = v - v_{peak} \tag{11}$$

However, even though transformations between images are dominantly translational, there usually is significant rotational and scaling as well, and therefore restricting the transformation set to translation is obstructive to precise georegistration. So by extending the concept of correlation surface super-imposition to incorporate a richer motion-model like affine, 'position-dependent' transforms like rotation, scaling and shear are included in the set of possible transformations. Once again the goal is to maximize the sum of the center position on all the correlation surfaces, only this time transformation of the correlation surfaces is not position independent. Each correlation surface, by virtue of the feature point around which it is centered, may have a different transformation associated with it. This transformation would depend on the absolute position of the element on the correlation surface rather than with its relative position as the affine set of transformations is not location invariant. An affine transform may be described by the six parameters specified in Equation 3 and 4. The objective then, is to find such a state of transformation parameters for the correlation surfaces that would maximize the sum of the pixel values at the original feature point locations corresponding to each surface. The affine parameters are estimated by directly applying transformations to the correlation surfaces. Figure 6 shows the correlation surfaces before and after transformation. It can be observed that the positions of the center of correlation surfaces i.e. $\eta(0,0)$ remain fixed in both images. In practice, window sizes are taken to be odd, and the sum of four pixel values around $\eta_i(0,0)$ are considered. The sum of the surfaces is once again expressed as in 9, where η_1 is the set of n affine-transformed correlation surfaces. This time the relationship between (u',v') and (u,v) is defined as,

$$x_f - u' = a_1(x_f - u) + a_3(x_f - u) + a_5 \tag{12}$$

$$y_f - v' = a_2(y_f - v) + a_4(y_f - v) + a_6 \tag{13}$$

and a search is performed over a so as to maximize. Thus, the function to be maximized is,

$$F(\mathbf{a}) = \eta(0,0). \tag{14}$$

In a sense, the correlation surfaces are affine-bound together to recover the most consistent set of peaks. It should be noted that the range of the correlation surface depends on the search window size, which in turn depends on the size of the orthorectified image. This search is performed over a pyramid, and alignment recovered is propagated to the next level. The recovered alignment is also applied to feature points as they are propagated to a higher resolution level, so that correlation may be performed at each level of the pyramid. The benefit of using this hierarchical approach is that it improves computational efficiency and avoids the aliasing of high spatial frequency components that require large displacements. To visualize the entire process, consider a feature point $I_{video}(x_f, y_f)$. A patch of nine by nine pixels around $I_{video}(x_f, y_f)$ is correlated with a fifteen by fifteen pixel search window around $I_{ref}(x_f, y_f)$ to yield the correlation surface η_f. Each element $\eta_f(u, v)$ on the correlation surface is treated as a similarity measure for the vector from $I_{video}(x_f, y_f)$, to the point $I_{video}(x_f + u, y_f + v)$. When the search is performed over the affine parameters, the affine transformation is applied there are n correlation surfaces and each surface is transformed according to the absolute position of the feature point around which it was centered. The task is to find the six affine parameters such that the sum of the values at the center block in each correlation surface (or F) is maximized. Once alignment is recovered it is propagated to a higher resolution level and correlation surfaces are computed around the feature points again and the process is repeated. Maximization of F is achieved by a Quasi-Newton optimization procedure, using a finite-difference computation of the relevant derivatives. Because the positional information is maintained, every iteration places a set of points of the correlation surface onto the feature point around which each surface was initially centered. As the optimization progresses further the method moves towards a consistent set of peaks. Transformations were propagated through the three bottom levels of a Gaussian Pyramid to ensure that large displacements are smoothly captured. It is worth noting that as the set of consistent correlation peaks are being transformed to the feature point locations of the orthorectified image, it is the actually the inverse affine transformation that is computed.

The advantage of maximizing in the process detailed is three-fold. Firstly, by maintaining a 'continuous' correlation surface (rather than thresholding for peaks and performing consistency measurement on them) the most consistent set of peaks in the correlation surface is naturally retrieved. This avoids thresholding and loss of image correlation details. Secondly, by considering surfaces, relative strengths of peaks are maintained: a stronger peak holds greater weight on the overall maximization process. Thirdly, the algorithm returns the optimal affine fit, without the need for an extra consistency step. In effect, the consistency and local alignment process are seamlessly merged into one coherent module.

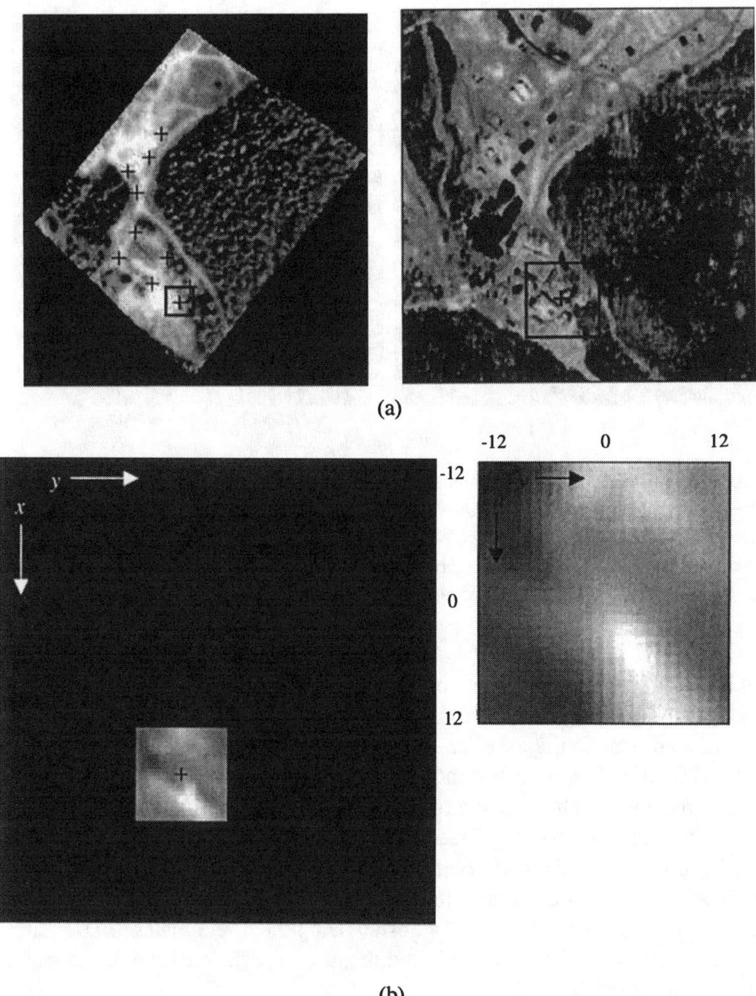

Fig. 5. The coordinate systems of correlation surfaces (a) The orthorectified image and the cropped reference image. The smaller window in the orthorectified image is the feature patch and the larger window in the cropped reference image is the corresponding search area. (b) The absolute coordinate system for the resulting correlation. The coordinate system is (x,y) of the original image. The black '+' indicates the position of (x_f, y_f). (c) The relative coordinate system, (u, v) defining distance from the feature point (x_f, y_f) shown as the black '+' in (b). The '⊗' shows the position of the peak in the correlation surface. The lack of any distinct peak should be noted, a typical phenomenon due to the differences between reference and video data.

Direction Registration

Once the Reference Image and the Aerial Video Frame have been aligned through feature-based registration, a direct hierarchical registration method is employed to

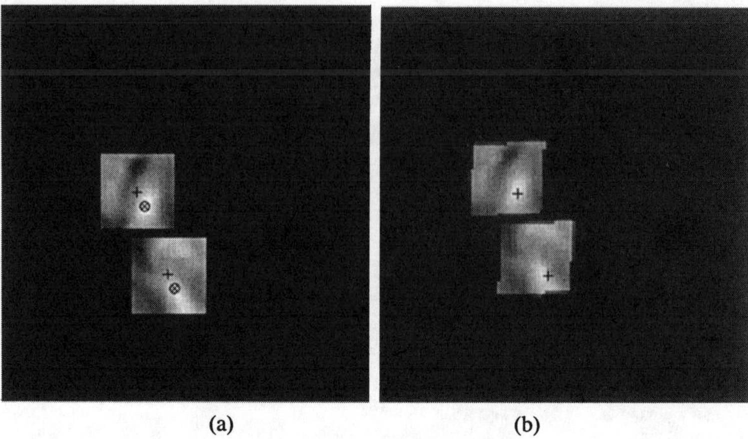

Fig. 6. Absolute position of correlation surfaces before and after transformation. (a) The '+'s mark the positions of the feature points. Two correlation surfaces are shown for illustrative purposes as the other overlap. The '$\otimes$' indicates the position of the dominant peak. (b) The correlation surfaces are transformed according to their absolute positions such that values at the '+'s are maximized. The position of the '+' remains the same in both (a) and (b).

provide a final adjustment. Feature based methods characteristically have a 'window' alignment, thereby losing information in the process of registration. To ensure that the whole image information is used, an affine direct registration is applied as proposed in [3] and [20]. The final transformation between the Aerial Image and the Cropped Reference Image is then the product of the affine transforms recovered from the Local Feature Match and this direct registration. As a general rule of minimization, the closer the initial estimate is to the true solution the more reliable the minimization process will be. The solution obtained after the feature-based alignment provides a close approximation to the answer, it is then adjusted using the direct method. To ensure that only a fine adjustment of the feature based method is performed the direct method is implemented for a single level.

3 Sensor Update

So far two-dimensional registration of the ortho-rectified Aerial Image and the Cropped Reference Image has been achieved. The registration is performed in the orthographic viewing space, providing six affine parameters. Using this 2D alignment, it is possible to assign 3D geodetic coordinates to every pixel by simple pixel-to-pixel correspondence from the Reference Image. The final objective of this paper is to recover the adjustment to the sensor model parameters to affect alignment. However, in order to recover the sensor model's nine parameters further processing is required. It is observed that there exists no unique solution (state of sensor parameters) corresponding to any given affine transformation. The following three parameter pairs, in particular, create an infinite space of solutions: (1) The camera focal length and the

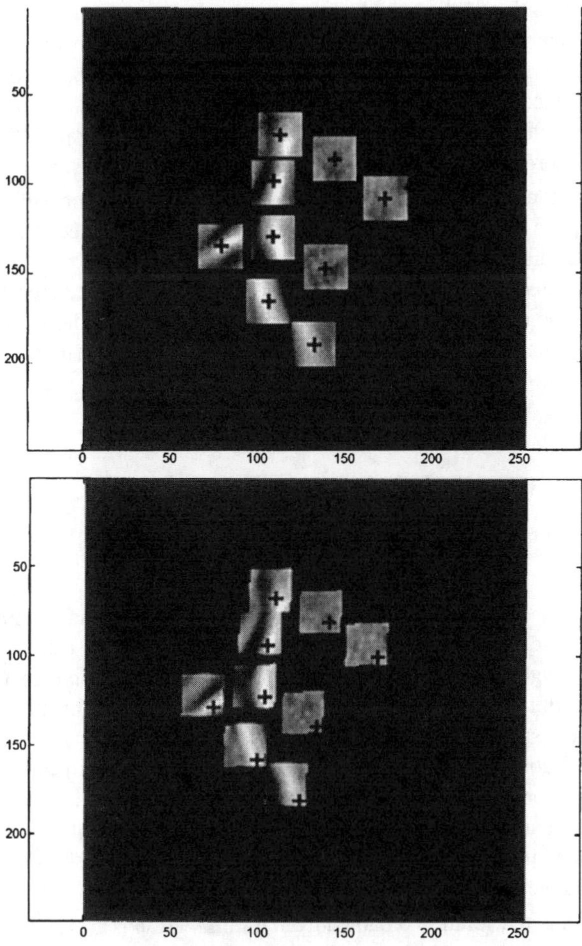

Fig. 7. The top figure show the positions of nine of the twenty feature points marked by '+'s and their correlation surfaces centered at each feature point for the second level of the matching pyramid. The bottom figure shows the results after alignment. It should be noted that the black '+'s do not change their position. After the iterations the correlation surfaces are all positioned so as to maximize the values at the feature points. It is worthwhile to note that the three surfaces in the tree textured area were 'ignored'. For illustration purposes this diagram displays only those surfaces that do not overlap, at the second level of a four level pyramid.

vehicle height, (2) The camera scan angle and the vehicle heading, and (3) The camera elevation and the vehicle pitch. Each one can have an approximate canceling effect, as the other in the pair, on the image captured. For instance, take the mutual effect of the camera focal length and the height of the camera. Increasing the vehicle height or decreasing the camera focal length achieves an equivalent effect of the captured image. To recover a plausible update of the sensor information two constraints are applied. Firstly, covariance information for each parameter is used while estimating the accurate updates of the sensor parameters and secondly the constraint of least change is applied in the form of a distance measure from the original sensor parameters state.

To recover the sensor adjustments, point correspondences are established between the Aerial Image and the Reference Image using the recovered 2D transformation. The Euclidean distance between those points are then minimized by searching over the nine parameters of the sensor model applying the constraints mentioned. As mentioned earlier the error function is critical to obtaining the fundamentally meaningful adjustments in the sensor geometry. The error function employed here was

$$E = \kappa_1 \Lambda(\mathrm{s}, I_{ref}, I_{video}) + \kappa_2 \Psi(\mathrm{s}, \mathrm{s}') \tag{15}$$

where

$$\mathrm{s} = \begin{bmatrix} s_1 & s_2 & s_3 & s_4 & s_5 & s_6 & s_7 & s_8 & s_9 \end{bmatrix} \tag{16}$$

are the nine sensor parameters, vehicle longitude, vehicle latitude, vehicle height, vehicle roll, vehicle pitch, vehicle heading, camera scan angle, camera elevation and camera focal length, s' is the initial telemetry state. $\boldsymbol{\Lambda}$ gives the Euclidean distance between the point correspondences of the two images using the current estimate of sensor parameters. The original set of points is back-projected onto the image plane, and a search is conducted to find a state of s that maps the projections of the points to their matches on the ground. $\boldsymbol{\Psi}$ calculates the weighted Euclidean distance of each adjusted sensor parameter from the initial telemetry data (weighted on the basis of the covariance found in the telemetry). κ_1 and κ_2 are constants whose sum equal one, used to assign a relative importance to the two constraints.

To ensure that the solution obtained from minimization is accurate two safe-checks are employed. First, a least change constraint is placed to ensure that the solution is realistically close to the original values. Second, the covariances provided in the telemetry are used to weight the minimization process to provide unique solutions. To manually calculate the analytical expression for the Jacobian required by the optimization would probably take the better part of a week, so symbolic toolboxes of any commercial mathematics software package can be used to generate the expressions. The expressions would be the expanded form of

$$\overrightarrow{X}_{camera} = \Pi_t \overrightarrow{X}_{world}, \tag{17}$$

where the coordinate transformation matrix Π_t is

$$\Pi_t = \begin{bmatrix} \cos\omega & 0 & -\sin\omega & 0 \\ 0 & 1 & 0 & 0 \\ \sin\omega & 0 & \cos\omega & 0 \\ 0 & 0 & 0 & 1 \end{bmatrix} \begin{bmatrix} \cos\tau & -\sin\tau & 0 & 0 \\ \sin\tau & \cos\tau & 0 & 0 \\ 0 & 0 & 1 & 0 \\ 0 & 0 & 0 & 1 \end{bmatrix}$$

$$\cdots \begin{bmatrix} \cos\phi & 0 & -\sin\phi & 0 \\ 0 & 1 & 0 & 0 \\ \sin\phi & 0 & \cos\phi & 0 \\ 0 & 0 & 0 & 1 \end{bmatrix} \begin{bmatrix} 1 & 0 & 0 & 0 \\ 0 & \cos\beta & \sin\beta & 0 \\ 0 & -\sin\beta & \cos\beta & 0 \\ 0 & 0 & 0 & 1 \end{bmatrix}$$

$$\cdots \begin{bmatrix} \cos\alpha & -\sin\alpha & 0 & 0 \\ \sin\alpha & \cos\alpha & 0 & 0 \\ 0 & 0 & 1 & 0 \\ 0 & 0 & 0 & 1 \end{bmatrix} \begin{bmatrix} 1 & 0 & 0 & \Delta T_x \\ 0 & 1 & 0 & \Delta T_y \\ 0 & 0 & 1 & \Delta T_z \\ 0 & 0 & 0 & 1 \end{bmatrix}, \tag{18}$$

or more concisely,

$$\vec{X}_{camera} = G_y G_z R_y R_x R_z T \vec{X}_{world}, \tag{19}$$

where G_y is a rotation matrix in terms of the camera elevation angle ω, G_z is a rotation matrix in terms of the camera scan angle τ, R_y is a rotation matrix in terms of the vehicle pitch angle ϕ, R_x is a rotation matrix in terms of the vehicle roll angle β, R_z is a rotation matrix in terms of the vehicle heading angle α, T is the translation matrix derived from the vehicle latitude, longitude and height. Details of converting vehicle longitude and latitude to meter distances from a reference point can be found using many cartographic texts. Here it is assumed that the vehicle displacements ΔT_x, ΔT_y and ΔT_z have been computed.

4 Results

To demonstrate the algorithm presented in this paper alignment for examples are presented in this section. Despite the substantial illumination change to the extent of contrast reversal (for watery areas), examination of the results shows a significant improvement on the initial estimate. Figure 9, 10 and 11 show the initial Video Frame and Reference Imagery before and after registration. It should be noted that the image sizes are upto 1500x1500 pixels, and figures are not to scale. The misalignments are therefore appear to be scaled down as well. Visual inspection reveals a misalignment after ortho-rectification of the Video Frame using the telemetry and sensor model. Attempts at minimizing this misalignment using brightness consistency constraints fails, but with the proposed Correlation Surface Binding Algorithm proposed in this paper, accurate alignment is achieved. Figure 12 provides further examples of correct registration. White circles are marked on the top two images to highlight the corrected positions of features in the Aerial Video Frame.

The portion of the image set on which the algorithm presented did not perform accurately, were of three types. The first type was images without any features at all, like images of textured areas of trees. As there were no real features to use as constraints, the performance on these images was sub-par. The second problem faced was the aperture problem where features present were linear, and thus only a single dimensional constraint could be retrieved from them. The most convincing method of addressing both these issue is using some form of bundle adjustment as was used in [6] and [31]. These methods were not used in this work since only video key-frames with little or no overlap were available. The last problem faced was that of occlusion by vehicle parts like tires and wings. This was addressed by calculating the fixed positions of the vehicle parts with respect to the camera in terms of the camera parameters (camera elevation angle, camera scan angle, and camera focal length). The portion of the image is then ignored or if it happened to cover too much of the image space, it is summarily rejected.

The results yielded a pre-registration average error of 39.56 pixels and a post-registration average of error 8.02 pixels per frame. As ground truth was not available to assess the error automatically, manual measurement was performed per frame. The results on a 30 key-frame clip is shown in Figure 8. The key-frames in the clip contained adequate visual context to allow single frame registration. Linear features were encountered causing some of the higher average errors reported.

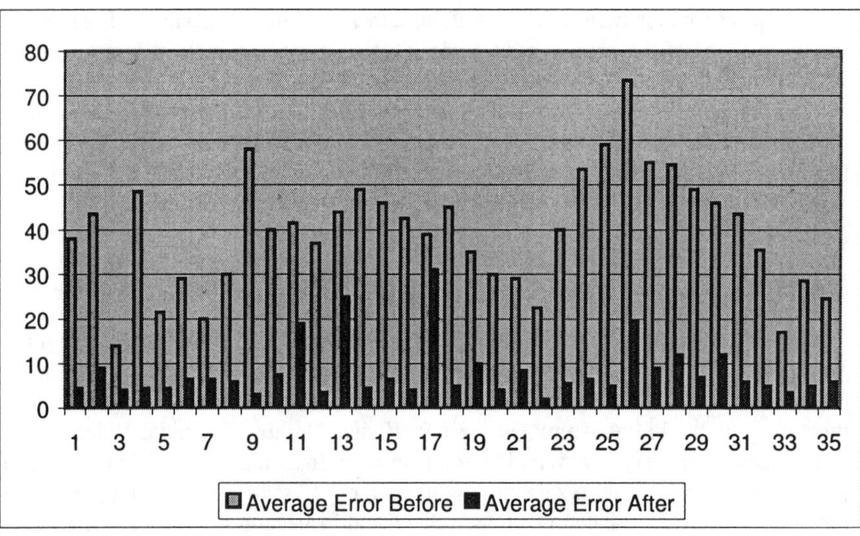

Fig. 8. Average error improvements over a 30 key-frame clip. Frame numbers are numbered along the horizontal axis, while errors in terms of number of pixels are specified along the vertical axis.

5 Conclusion

The objective of this paper was to present an algorithm that robustly aligns an Aerial Video Image to an Area Reference Image and plausibly updates the sensor model parameters, given noisy telemetry information along with elevation values for the Area reference image. The major problems tackled here were rectifying the images

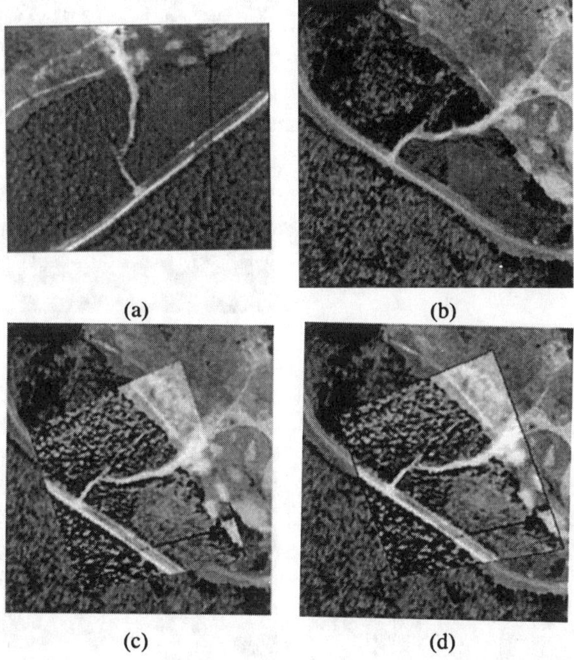

Fig. 9. Geo-registration results: (a) Aerial video frame; (b) Cropped area of reference image; (c) Orthorectified video frame placed upon cropped reference image; (d) Gross misalignment by parametric direct registration over a four level pyramid (using an affine transform); (e) Registration by feature linking.

by bringing them into a common viewing space, geodetic assignment for aerial video pixels, and sensor model parameter adjustment. Various forms of distortions were tackled, adjusting for illumination, compensating for texture variation, handling clouds and occlusion by vehicle parts. To achieve registration, the images are equalized and rectified into an orthographic viewing space, after which Gabor features are extracted and used to generate a normalized correlation surface per feature point. The hierarchical affine-based feature alignment provides a robust coarse registration process with outlier rejection, followed by fine alignment using a direct method. The sensor parameters are then adjusted using the affine transformation recovered and the distance of the solution from the original telemetry information. It is to be expected that the sensor data will improve with the forward march of technology, bringing with it the possibilities of more sophisticated models for the georegistration problem. Any improvement in the elevation data in particular would allow more confident use of three-dimensional information and matching. Future directions of the work include solving the initial alignment robustly in the perspective viewing space using more realistic rendering, and performing registration without continuous telemetry information.

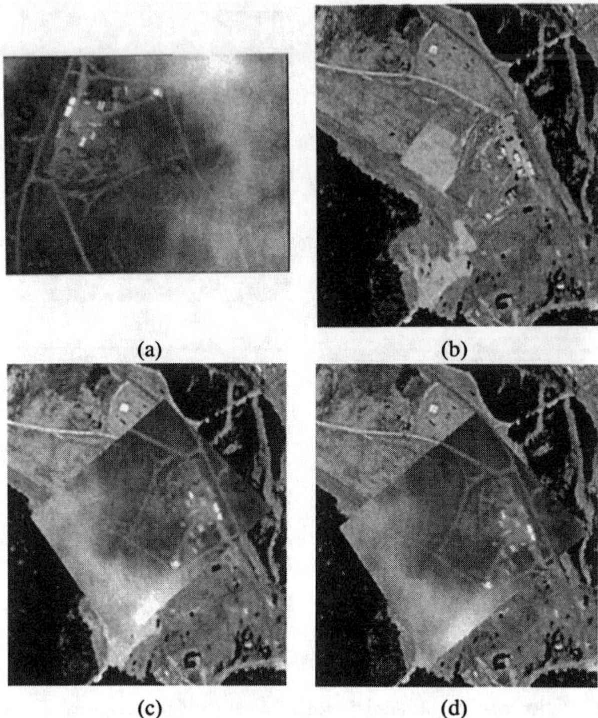

Fig. 10. Geo-registration results: (a) Aerial video frame; (b) Cropped area of reference image; (c) Orthorectified video frame placed upon cropped reference image; (d)Gross misalignment by parametric direct registration over a four level pyramid (using an affine transform); (e) Registration by feature linking.

References

1. P.Anandan,"A Computational Framework and an Algorithm for the Measurement of Visual motion", *International Journal of Computer Vision*, vol.2, pp. 283-310, 1989.
2. C.Baird and M. Abrarnson, "A Comparison of Several Digital Map-Aided Navigation Techniques", *Proc. IEEE Position Location and Navigation Symposium*, pp. 294-300, 1984.
3. J.Bergen, P. Anandan, K. Hanna, R. Hingorani, "Hierarchical model-based motion estimation", *Proc. European Conference on Computer Vision*, pp. 237-252, 1992.
4. L. Brown, "A Survey of Image Registration Techniques", *ACM Computing Surveys* , 24(4), pp. 325-376, 1992.
5. Y. Bresler, S. J. Merhav, "On-line Vehicle Motion Estimation from Visual Terrain Information Part II: Ground Velocity and Position Estimation" , *IEEE Trans. Aerospace and Electronic System*, 22(5), pp. 588-603, 1986.
6. R. Cannata, M. Shah, S. Blask, J. Van Workum, "Autonomous Video Registration Using Sensor Model Parameter Adjustments", *Applied Imagery Pattern Recognition Workshop,* 2000.
7. P. Curran, *"Principles of Remote Sensing"*, Longman Group Limited, 1985.

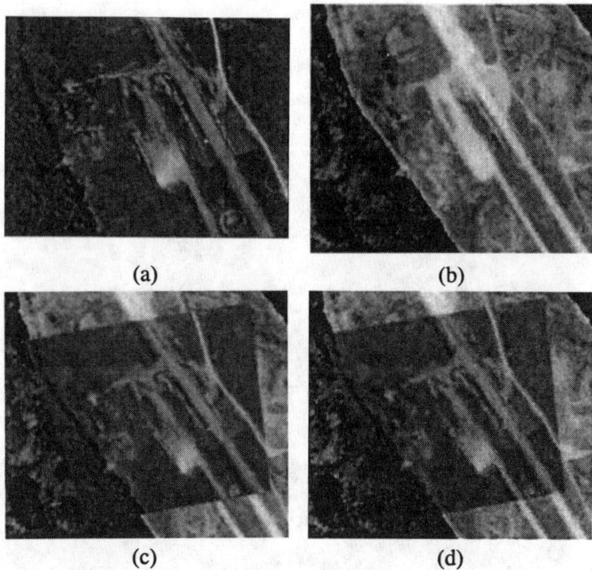

Fig. 11. Geo-registration results: (a) Aerial video frame; (b) Cropped area of reference image; (c) Orthorectified video frame placed upon cropped reference image. (d) Gross misalignment by parametric direct registration over a four level pyramid (using an affine transform); (e) Registration by feature linking.

8. J. Foley, A. van Dam, S. Feiner, J. Highes, *"Computer Graphics, Principles and Practices"*, Addison-Wesley, 1990.
9. D. Gabor, "Theory of Communications", *IEEE Communications*, No. 26, 1946, pp. 429-459.
10. J. Golden, "Terrain Contour Matching (TERCOM): A Cruise Missile Guidance Aid", *Proc. Image Processing Missile Guidance*, vol. 238, pp. 10-18, 1980.
11. V. Govindu and C. Shekar, "Alignment Using Distributions of Local Geometric Properties", *IEEE Transactions on Pattern Analysis and Machine Intelligence*, 21(10), pp. 1031-1043, 1999.
12. K. Hanna, H. Sawhney, R. Kumar, Y. Guo, S. Samarasekara, "Annotation of video by alignment to reference imagery", *IEEE International Conference on Multimedia Computing and Systems*, vol.1, pp. 38 - 43, 1999.
13. B. Horn, B. Schunk, "Determining Optical Flow", *Artificial Intelligence*, vol. 17, pp. 185-203, 1981.
14. S. Hsu, "Geocoded Terrestrial Mosaics Using Pose Sensors and Video Registration", *Computer Vision and Pattern Recognition*, 2001. vol. 1, pp. 834 -841, 2001.
15. http://ams.egeo.sai.jrc.it/eurostat/Lot16-SUPCOM95/node1.html
16. M. Irani, P. Anandan, "Robust Multi-Sensor Image Alignment", *International Conference on Computer Vision*, 1998.
17. B. Kamgar-Parsi, J.Jones, A.Rosenfeld, "Registration of Multiple Overlapping Range Images: Scenes without Distinctive features", *Computer Vision and Pattern Recognition*, pp. 282-290, 1989.

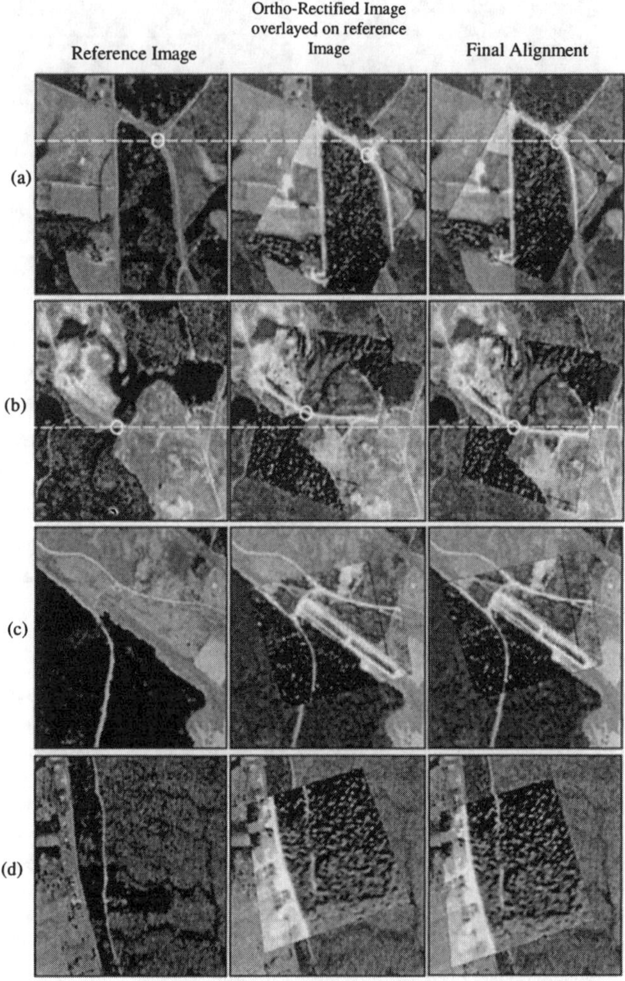

Fig. 12. (a)-(d) The leftmost image is the cropped reference image, the middle image is the orthorectified image overlayed onto the reference image, and the rightmost image is the final registered image. The white circles highlight initially misaligned features.

18. R. Kumar, H. Sawhney, J. Asmuth, A. Pope, and S. Hsu, "Registration of Video to Georeferenced Imagery", *Fourteenth International Conference on Pattern Recognition* vol. 2. pp.1393-1400, 1998.
19. B.Lucas and T.Kanade."An Iterative Image Registration Technique with an Application to Stereo Vision", *Proceedings of the 7th International Joint Conference on Artificial Intelligence,* pp. 674-679, 1981.
20. S. Mann and R. Picard,"Video Orbits of the Projective Group: A Simple Approach to Featureless Estimation of Parameters", *IEEE Transact. on Image Processing*, 6(9), pp. 1281 -1295, 1997.
21. S. J. Merhav, Y. Bresler, "On-line Vehicle Motion Estimation from Visual Terrain Information Part I: Recursive Image Registration", *IEEE Trans. Aerospace and Electronic System,* 22(5), pp. 583-587, 1986.
22. J. Le Moigne, N. Netanyahu, J. Masek, D. Mount, S. Goward, M. Honzak, "Geo-Registration of Landsat Data by Robust Matching of Wavelet Features", *Proc. Geoscience and Remote Sensing Symposium, IGARSS* , vol.4, pp. 1610-1612, 2000.
23. J. Nocedal, S. Wright, *"Numerical Optimization"*, Springer-Verlag, 1999.
24. J. Rodriquez, J. Aggarwal, "Matching Aerial Images to 3D Terrain Maps", *IEEE PAMI*, 12(12), pp. 1138-1149, 1990.
25. D.-G. Sim, S.-Y. Jeong, R.-H. Park, R.-C. Kim, S. Lee, I. Kim, "Navigation Parameter Estimation from Sequential Aerial Images". *Proc. International Conference on Image Processing*, vol.2, pp. 629-632, 1996.
26. D-G. Sim, R-H Park, R-C. Kim, S. U. Lee, I-C. Kim, "Integrated Position Estimation Using Aerial Image Sequences", *IEEE Transactions on Pattern Analysis and Machine Intelligence*, 24(1), pp. 1-18, 2002.
27. R. Szeliski, "Image Mosaicing for Tele-Reality Applications", *IEEE Workshop on Applications of Computer Vision*, pp. 44-53, 1994.
28. Y. Sheikh, S. Khan, M. Shah, R. Cannata, "Geodetic Alignment of Aerial Video Frames", in *Video Registration*, Video Computing Series, KLUWER Academic Publisher, 2003.
29. R. Szeliski, H. Shum, "Creating Full View Panoramic Image Mosaics and Environment Maps", *Computer Graphics Proceedings, SIGGRAPH*, pp. 252-258, 1997.
30. P. Viola and W.M. Wells, "Alignment by Maximization of Mutual Information.", *International Journal of Computer Vision*, 24(2) pp. 134-154, 1997.
31. R. Wildes, D. Hirvonen, S. Hsu, R. Kumar, W. Lehman, B. Matei, W.-Y. Zhao "Video Registration: Algorithm and Quantitative Evaluation" , *Proc. International Conference on Computer Vision*, Vol. 2, pp. 343 -350, 2001.
32. Q. Zheng and R. Chellappa. "A Computational Vision Approach to Image Registration", *IEEE Transactions on Image Processing* 2(3), pp. 311 326, 1993.

Acquisition of a Predictive Markov Model using Object Tracking and Correspondence in Geospatial Video Surveillance Networks

Christopher Jaynes

Dept. of Computer Science
University of Kentucky
jaynes@cs.uky.edu

ABSTRACT

Multi-view surveillance in large and complex environments involves a large number of cameras spread over a wide area, often with no overlapping fields of view. This paper explores how global models of object motion under such conditions can be automatically acquired. These models can then be used to predict subject motion through the network and, in some cases, derive geospatial constraints on the position of network sensors.

We introduce the *topological flow model,* a Hidden Markov description of relative camera position and traffic flow, where each state is a camera view and edges between states represent subject paths between cameras. Edge weights represent the transition probability that a subject seen in a particular view will next appear in a second view. Tracking and matching of subjects that move from one view to the next provides the initial estimate of the multi-view topology. Transition probabilities are computed so as to maximize the posterior probability given observed subject motions. We present initial experimental results on a six-camera network that tracks vehicles over a large geospatial area.

Keywords— Markov model, Motion analysis, Object matching, Tracking, Geospatial surveillance.

1. INTRODUCTION

Automated visual surveillance within complex environments involves the use of multiple stationary cameras that are potentially deployed over a very large geospatial area. We refer to these types of camera networks as geospatial surveillance networks in order to distinguish the unique problems associated with this domain from traditional single and multi-view surveillance scenarios. Geospatial surveillance requires that cameras are placed to facilitate surveillance goals for a particular application such as coverage maximization of well-trafficked areas (i.e. a road intersection), are placed to capture specific views of a subject or region (i.e. the entrance to an office or parking garage), or are deployed to support other computer vision tasks such as three-dimensional model acquisition. Placement constraints such as these, along with the significant potential for occlusion between sensors due to wide separation, often precludes overlapping fields of view and explicit camera-to-camera calibration of the individual devices. Given these problems, several researchers are addressing the unique computer vision issues associated with large numbers of cameras, spread over large areas, with unknown calibration [2,6,19,21].

This paper explores how a globally coherent model that describes both the motion paths of tracked objects and the general topology of the sensor network can be automatically acquired. T*opological calibration* is the problem of recovering a probabilistic state transition description for objects moving within a surveillance network composed of stationary cameras. The topological model is related to camera position, characteristics of the environment such as the presence of occluding objects, and the characteristics of subject motion through the deployed network.

This work is motivated by the observation that traffic, in many domains, , is somewhat constrained to a set of fixed paths in the environment. Examples include, vehicle movement on urban roadways and crowds of people moving through an the hallways of a building. Over time, a model, based on these common paths, can be acquired and used to discover the topological relationship between cameras in the network.

Under these conditions we introduce a method to acquire a *topological flow model.* States in the model correspond to cameras in the surveillance network and edges between nodes i and j imply that subjects are capable of moving from camera view i to camera j without first being detected in a third view. Edge weights represent the probability that an subject currently being tracked in view i will next be seen in view j. We assume that object motion is a Hidden Markov process, that is, state transition probabilities are dependent upon the current state (camera) in which the object is currently being tracked and the underlying transition probabilities are only partially observable through a (potentially noisy) process that matches subjects as they move between views. The topological calibration problem, then, is to acquire this model automatically by tracking subjects in all cameras, matching subjects across views to detect transitions, and update transition probabilities so as to maximize the posterior probabilities given the observed data.

Once acquired, models can be used to analyze traffic behavior, predict subject paths, and potentially detect subject motions that are anomalous. In interactive video surveillance applications, the topological model can assist users who are monitoring the motion of subjects in a large surveillance network. As a subject exits one view, the acquired topological description can be used to present a set of views to the user based on expected subject paths. Acquiring such a model by hand, particularly in large networks placed over wide areas is infeasible.

We speculate that the topological model can be used to derive geospatial constraints on the position of cameras in the network. Given the network topology and observed object transition times, the relative distance between sensors can be established. When used in conjunction with a DEM and/or vector data describing roadways and cultural features, the position of cameras may be constrained [3]. Perhaps most importantly, other researchers have shown that predictive models such as the topological model we propose here be used to improve tracking and analyze subject motion for people tracking scenarios [15].

2. RELATED WORK

Tracking of motion from large-scale camera networks, potentially deployed over large geographic areas is a challenging problem. A number of different research efforts are addressing the problems related to wide-area surveillance including cooperative object tracking [5,21,23], multi-view object recognition [16], trajectory estimation [4], data fusion from widely separated sensors, particularly in urban environments [17,22], active control of cameras [2,16], and camera calibration from motion [12,17]. More recently, calibration, tracking , and matching subjects across views in large camera networks with potentially non-overlapping views has been addressed by [6,11,12] and recent research has shown that predictive models of motion in these scenarios are useful [15]. Primarily, however, research has focused on multi-view tracking and analysis in domains that assume significant overlap between views.

Topological calibration involves tracking of subject motion and matching these moving objects in widely separated views. These problems are the subject of significant research and are not the focus here. The reader is referred to [2] for an overview of object tracking and matching for video surveillance applications. Any tracking and matching algorithms that are known to be reliable for the domain under consideration can be used to acquire the topological model. The particular techniques used here are related to appearance based object recognition as well as background modeling using a mixture of Gaussians [9,20] .The matching algorithm presented here is similar to that used in a variety of vision tasks including face detection [7] and background modeling for segmentation [20].

The recovered topological flow model is based on traditional Hidden

Markov state representations. State transition probabilities are adjusted from the matched object transitions using the well known Baum-Welch equations so as to maximize the posterior likihood of the observed data [1]. Although there is work that uses statistical description of motion trajectories for motion analysis in a single view [14], our contribution is the recovery of a transition model that is globally consistent. The model can then predict when and where subjects are likely to appear in the calibrated network of cameras [11,15].

3. TECHNICAL DETAILS

Topological calibration involves three components. First, the network of cameras must be able to reliably track moving objects in their field of view. Secondly, an appearance model of tracked subjects must be extracted in each view. These models support matching of moving subjects in other views. Finally, cameras must communicate via a network that also supplies each camera with a global synchronization signal. Network communication is used to transmit appearance models from each camera to a central server for matching. Each appearance model transmitted to the server is accompanied by a timestamp that encodes the time that a subject enters and exits that particular view.

3.1. System Overview

The topological calibration system is composed of set of stationary surveillance cameras, a network, and a centralized server. Each surveillance camera is connected to a computer that performs local image processing operations. All computer-camera pairs are directly connected to a sever machine via a network. The server machine broadcasts a time signal to all machines so that a global clock is available to the system. The topological model is computed at the server by matching incoming object appearance descriptions from each view. The model can be updated as new cameras are added to the network, or significant changes in subject motions occur.

During calibration, each camera is responsible for tracking moving objects and deriving an appearance model for the target region in that view. For each tracked object, k, an object description vector is produced by the camera that includes the time interval in which it was observed, $[t_0, t_1]_k$, a camera identifier A, and a corresponding appearance model M_k. As each object leaves the view of a camera, this information is transmitted to the server for processing and potential update of the topological model.

The server compares each new object description vector to the set of vectors currently stored on an *unmatched objects list* for all objects whose exit time t_1 is less than the entrance time t_0 of the object under consideration. This temporal constraint is important in reducing the number of false positive matches and is possible because a global time clock is available via the network and cameras are assumed to be non-overlapping. Known network delays can be taken into account at this phase to adjust the entrance/exit time inequalities.

Each comparison between an incoming object description vector, V_A, and a vector on the unmatched list, V_B, results in one of three possible outcomes. First, if the similarity between V_A and V_B is greater than the similarity between V_A and all other unmatched objects and exceeds a predefined threshold, T_H, the two objects have been successfully matched across views. In this case, V_B is removed from the unmatched list, V_A is added to the list, and the topological model is updated to reflect the increased likelihood of object transition between cameras A and B.

If the similarity between two the objects V_A and V_B exceeds threshold T_L but is not greater than the similarity between all other objects on the unmatched list, then object V_B is eliminated from the unmatched objects list, V_A is discarded, and neither object will be used to update the topological model. This lower bound threshold, T_L, is used to eliminate objects from consideration that were similar, but not matched to, an incoming object. This threshold must be less than the matching threshold T_H. Elimination of objects under these circumstances is used to increase the separability of objects currently on the unmatched list for future matching. Although removal of unmatched objects will cause potential correspondences to be missed and increase the time required to converge to a calibrated network, empirical results have shown that culling of the unmatched objects list in this way increases the accuracy of the recovered model.

If the similarity score between V_A and all unmatched objects does not exceed T_L, the new object description vector is simply added to the unmatched objects list. The object is now available to be matched against any new object observed in other cameras in the network. Finally, the exit time, t_l, for remaining elements on the unmatched object list is subtracted from the current clock time to compute how long each object has been unmatched. Objects whose time on the list exceeds a maximum time threshold are discarded and the transition probability between the camera corresponding to the expired object and a global "exit" state is increased. The expiration time threshold is related to the maximum transition time between any two cameras in the surveillance network and was set to be 2 minutes for the results discussed here.

3.2. Tracking of Object Motion and Model Extraction

Object tracking is based on a straightforward foreground extraction and inter-frame matching algorithm, guided by a Kalman filter. Foreground regions are detected independently in each frame as pixels that do not adhere to background model. These regions are then matched according to proximity and color similarity measures to derive an image trajectory for each moving object in the scene.

An adaptive background model is constructed in each camera by analyzing the color distribution of each pixel over a window of time T. Each set of values for a single pixel, $p(i,j)_t=x_t,\ t=t_0,\ldots,t_0+T$, produce both a pixel mean $p(i,j)_\mu$ and variance, $p(i,j)_\sigma$. These values correspond to a Gaussian that best fits the observed values over time period T. In order to adapt the model, a sliding window of size T constantly updates the samples for each pixel and the corresponding background mean and variance values. Figure 1a depicts the mean pixel values assigned to the background model for a single camera over a 100 second interval. The corresponding regions in motion are shown in Figure 1b. Although there are more sophisticated approaches to background modeling and foreground segmentation (i.e. use of multi-modal distributions

(a) (b)

Figure 1: Tracking vehicles using background modeling and foreground segmentation. (a) Adaptive background model constructed using a 5 min. sliding window. Pixel intensity encodes mean color values that have been converted to intensity. (b) Variance of pixel color values in the background model. (c) Pixels that are more than two standard deviations from the model are grouped into regions based on proximity and color uniformity. Bounding regions show vehicles tracked in a single frame.

[10,20]), a single adaptive Gaussian model is sufficient for this work.

Given a background model, a frame-by-frame process detects pixels that are greater than two standard deviations from the current background model. Pixels labeled as foreground are then clustered into connected component blobs for analysis.

For each foreground blob, pixels are converted to YUV space and only pixel chrominance values are retained. We discard intensity in order to increase the robustness of matching across views that may vary in intensity due to dramatically different lighting conditions. Two blobs are merged into a single region if they are within $S/2$ pixels of one another, where S is the longest vertical or horizontal length of either blob and the mean color value of both the u and v components are within a predefined threshold.

These colored regions are then tracked in subsequent frames by searching a local neighborhood that is biased by a standard Kalman filter [13]. Moving mean and variance of the chromaticity values are computed from the current frame using all previous frames in which the object was tracked. Regions are matched in subsequent frames if the center of mass of the two regions is within the neighborhood defined by the Kalman filter and the mean chromaticity is within two standard deviations of the region to be matched. Tracking through occlusions is accomplished using a predictive look-ahead over k frames, guided by the Kalman filter [13]. If an tracked object is no longer observed, and cannot be matched to a new region emerging from an occlusion or exits the image plane, the current time is stored and an appearance model for the tracked motion is constructed. Figure 2 depicts several frames of a tracking sequence in which a yellow taxi cab enters the camera view and exits the scene after 47 frames.

Figure 2: Example of vehicle tracking. Black lines denote closed region of pixels being tracked (other objects being tracked are not denoted for visualization purposes). (a) Police vehicle being tracked as it enters the scene. (b) Tracker continues to match moving region through partial occlusions. (c) Object begins to exit scene after 46 frames. White line denotes computed object trajectory.

When an object is no longer tracked in the image, an appearance model is constructed from all chromatic samples of the subject gathered from the sequence of frame in which the object was tracked. These chromatic samples for T different frames define a 2D histogram in the UV color plane.

Two-dimensional, K Gaussians are fit to the 2D color histograms using a least squares technique. Each Gaussian model is steerable on the UV plane. That is, we determine the mean, variance, and planar rotation matrix, $\mathbf{R}$, for each mode of the model that best fits the observed data.

The probability that a target pixel, from the tracked region over all T frames, will has a chromaticity vector **X**, is estimated as:

$$\Pr(\mathbf{X}) = \sum_{i=1}^{K} \left[\frac{a_j}{(2\pi)^{\frac{m}{2}} |\Sigma_i|^{1/2}} e^{-\frac{1}{2}(\mathbf{X}-\mu_i)^T \Sigma_i^{-1} (\mathbf{X}-\mu_i)} \right] \mathbf{R}_i \qquad \text{(Eq. 1)}$$

Where a_j is the mode weight, μ_i is the mean, Σ_i is the covariance of the ith distribution, and $\mathbf{R}_i$ is a planar rotation matrix. Our method for modeling the statistical distribution of chromaticity values is similar to that of other approaches based on the standard mixture-of-Gaussians approach. However, we model the planar rotation of each mode to increase the flexibility of the model. Early experiments show that, for many scenarios, these extra degrees of freedom allow more accurate models but may be more sensitive to error for less saturated objects. For the work here, we track vehicles that tend to be quite saturated with specific color modes and the model is appropriate.

Given a chromatic distribution corresponding to an object trajectory, an offline EM algorithm estimates the Gaussian mixture model [1] using K distributions. The value of K should exceed the number of significant

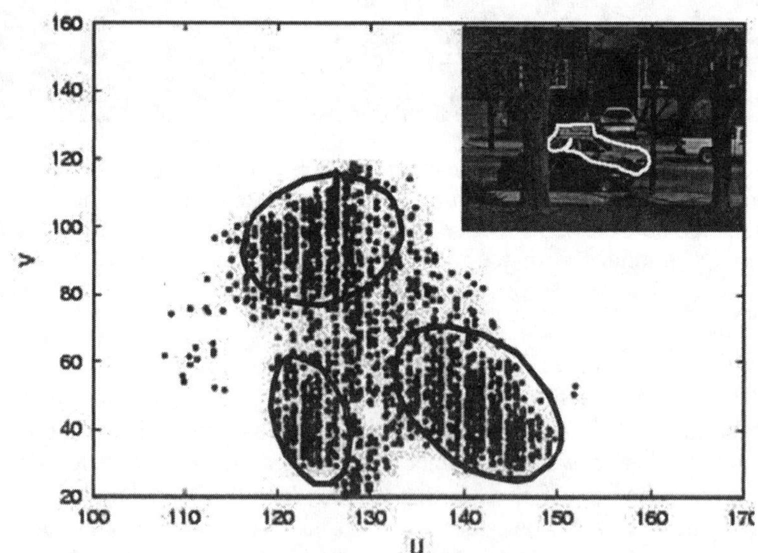

Figure 3: Two-dimensional Color histogram in U,V space. Histogram is shown projected to the U,V plane and magnitudes are not shown for ease of understanding. Votes in histogram accumulated over 53 frames while tracking the vehicle shown in inset in a single view.

components expected from the tracked objects and is related to the expected color variability both due to material properties of the object and lighting changes as the object moves through the scene. For the results shown here, K was fixed to six. The K distributions are ordered according to a_k/σ_k and the most significant N components of the fit components are used as the model of the object [20]. Although methods have been introduced to adaptively select the number of coherent modes dynamically based on a utility score, we fix N to 3 for all cameras and objects to simplify subsequent matching.

Figure 3 shows the set of points, projected to the (u,v) plane for a "taxi cab" object tracked in 53 frames (shown as inset in Figure 3). In this example, a total of 326,223 pixels samples were collected over the 53 frames in which the object was tracked.

Once an appearance model is estimated, an object description vector for the tracking event is constructed. In addition to the Gaussian mixture model parameters, the object description vector includes a unique number that identifies the camera that tracked the object, as well as global times corresponding to the entrance and exit of the object in that view. The object description vector is matched against other descriptors produced by other cameras.

3.3. Subject Matching in Widely Separated Views

Subjects on the unmatched object list are compared by measuring the similarity between the two corresponding Gaussian mixture models at relative offsets in the U, V plane. Although appearance change between views is a complex function of lighting, viewpoint, and material properties, we simply model these changes as a 2D translation in the chromatic *UV*-space. As part of our future work, we are exploring methods to estimate the non-linear color transfer function that maps the gamut of one camera in the network to another. These transfer functions can then be applied to object appearance models in order to increase matching robustness. It should be noted that, in this domain, a tradeoff between increased false negative matches for decreased false positives is justified. That is, when matches occur, they should be accurate in order to guarantee that the topological model will reflect the true underlying transition probabilities. False negatives only mean that the model will take longer to acquire as observed transitions, that could have been used to update the model, are lost.

There are several options for comparing the similarity of two functions at many relative offsets. For our purposes, we use the normalized cross correlation score.

The similarity between two object appearance models is given by:

$$\arg\max u,v \left(\frac{\sum_{u,v}\left[m_1(u,v)-\overline{m}_{1,ij}\right]\left[m_2(u-i,v-j)-\overline{m}_2\right]}{\left\{\sum_{u,v}\left[m_1(u,v)-\overline{m}_{1,ij}\right]^2 \sum_{u,v}\left[m_1(u-i,v-j)-\overline{m}_2\right]\right\}^{1/2}} \right) \quad \text{(Eq.2)}$$

where $m_{1,i,j}$ is the mean of the stationary model over the region covered by the extent of m_2 at the current offset and m_2 is the mean of the second object model. The values for both $m_1(u,v)$ and $m_2(u,v)$ are given by the 2D chromatic mixture models of the form given by Equation 1.

A similarity score, given by Equation 2, is computed between each incoming model and the current set of object models on the unmatched objects list that conform to the temporal constraints. If the maximum similarity between a new model and all objects exceeds a predefined upper bound threshold, U_T, then a match has been found. In this case, the topological model is updated appropriately and the matched object is removed from the unmatched object list while the new object is added for potential matches in the future. In the case that the similarity score exceeds a lower bound threshold, L_T but does not exceed U_T, then the new model is considered to be too similar and not separable from existing objects on the unmatched object list. In this case the object is discarded from the server. Finally, if the similarity score does not exceed U_T for any pair of objects, the new object description vector is simply added to the unmatched objects list.

It should be noted that this straightforward matching scheme was designed to support development of the topological model. Although the matching scheme is sensitive to dramatic lighting changes between views, or significant viewpoint variation, the upper-bound threshold parameter can be set very high to reduce the number of false positive matches. This ensures that when possible, objects are matched and the topological model is updated appropriately while avoiding false positives that can lead to an incorrect model. In addition, the lower bound threshold can be adjusted to remove objects that may yield false positive matches in future comparisons by forcing separability (in terms of the similarity score) on the objects currently on the unmatched objects list. For the results shown here values of $U_T=0.9$ and $L_T=0.73$ were used.

Figure 4 shows a vehicle that was reliably matched between three views of a surveillance network. This example was taken from the experiment described in Section IV and camera views correspond to cameras 1, 2, 3 in Figure 6.

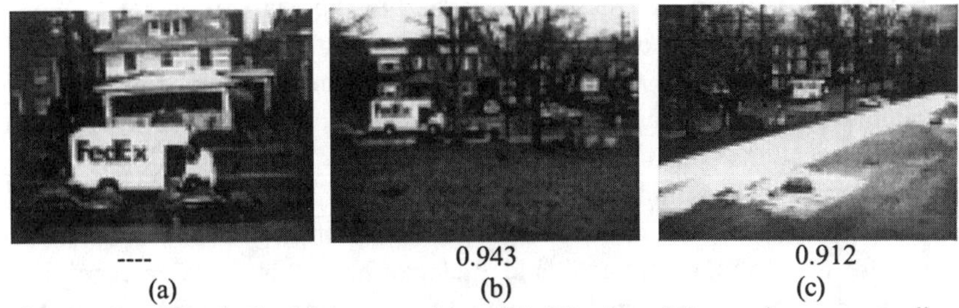

---- 0.943 0.912

(a) (b) (c)

Figure 4: Tracked object, a white "FedEx Truck" and corresponding similarity scores as the object moves through three different views in the camera network (a) First view of object, appearance model is constructed and transmitted to server for matching. (b) Same object, seen a few seconds later in a second view. Object model is again acquired and matched at the server with a score of 0.943. (c) Object is seen for a third time, and matched to the appearance model constructed in (b).

3.4. Topological Calibration

The general topological flow model is a Markov process whose set of nodes representing view locations that are connected via directed, weighted edges. Edge strengths represent the probability that an object in one view will next be seen in a connected view. Here we assume that the graph is fully connected. This assumption does not significantly impact the learned model because edges for which no transitions are observed will eventually converge to zero in the learning phase. Future work involves efficient ways to bias the initial topology of the graph for faster convergence to a set of valid transition probabilities.

The topological model is the 5-tuple, $(\Omega_x, \Omega_o, A, B, \rho)$. Where Ω_x is the set of possible states (camera views) that an object can reside in, Ω_o is the set of observations, A is a description of the transition probabilities, B is a description of prior observation probabilities, and ρ is the initial state distribution. For the results shown here, we assume the probability of observing a particular object in a given view, if it is indeed in that view is 0.95. Obviously, these probabilities are related to the accuracy of the particular tracking and recognition algorithms used by the system as well as the surveillance conditions such as potential for occlusions in particular views. In future work we expect to estimate these prior probabilities through direct analysis of the algorithm behavior in each view.

Transition probabilities are encoded in matrix form where entry A_{ij} represents the probability objects leaving view i are next seen in view j. Because subjects may exit the network at any time (i.e. turn down a path that is not monitored), an exit state, ε, is added to the Markov model. For *n* cameras, the goal of topological calibration, then, is to estimate the $(n+1)$ x$(n+1)$ transition matrix *A* that correctly reflects the statistical flow of traffic through the given cameras. Estimation of these transition probabilities from a set of observations *O* that maximizes $P(O \mid A)$.

Prior estimates of transition probabilities, A_0, are first set to equal $1/n$ where *n* is the number of cameras in the network. If prior estimates other than the uniform distribution are available, they can be set in A_0. For example, if two cameras are placed in such a way that no traffic should pass between them without first being observed in a third view, this can be encoded in A_0 as an initial transition probability of zero. In this work, we make no assumptions about the placement of cameras of the flow of traffic that they observe and show that uniform priors can lead to a useful topological model.

When a subject is matched, the current transition estimates are updated so that local transition probability expectation is maximized. If an object leaves view *i* and is subsequently matched in view *j*, the ith row of matrix *A* is updated. Each row is a prior distribution function, $p(X|i)$ that encodes the probability objects in view *i* will be observed in one of several states, represented by vector *X*. Local maximization of the posterior probability of the observation given the prior distribution is a classic problem and can be addressed using standard Expectation Maximization (EM) methods. For a description of the EM method as well as a description of the iterative application of the Baum-Welch equations for transition probabilities update given a set of observations (as is the case here), the reader is referred to [1].

4. EXPERIMENTAL RESULTS

The approach was demonstrated using a six-camera surveillance network. Cameras were placed over a few city blocks in an urban environment and

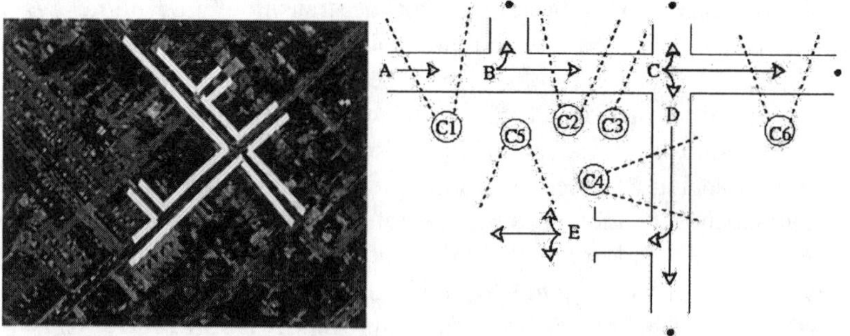

Figure 5: Diagram of experimental deployment. Six cameras were placed to monitor traffic flow through neighborhood streets. Circles denote placement of cameras and their corresponding dash-lines their field of view. Arrows show potential directions for traffic flow along city streets. Letters are referred to in text. See Figure 6 for images corresponding to each camera view.

oriented to observe vehicle traffic on several streets. All cameras in the experiment were low-power, digital color CCDs connected to a local computer via an IEEE 1394 interface. Each computer captured video at 15 frames per second, tracked objects and constructed the Gaussian mixture model online. A central server machine received object description vectors from all cameras, performed matching and updated the Markov model as object transitions are discovered.

In order to assist the reader in correctly interpreting the recovered transition probabilities, Figure 5 depicts the general layout of the experiment. An aerial view of the region, with an overlay depicting the streets under observation, is shown in Figure 5a. Figure 5b diagrams the camera locations (circles C1 through C6) and their general field of view with respect to the streets. Street layouts are shown and traffic directions are shown as arrows. Solid dots denote a potential exit from the surveillance network. Vehicles can enter the surveillance area along the one-way street A, from the intersection marked C, or along street D. Vehicles can exit the network at intersection B. This is a small side street that is rarely used by common traffic. Intersection C is a four-way junction. Vehicles may continue on the one-way street to be seen in camera C6, a left turn to exit the network, or turn right to pass along street D and be observed in camera C4. Point E on Figure 5b corresponds to the entrance to a parking lot. Vehicles entering the parking lot may exit the network by traveling straight for turning left. A turn right at E will lead to a section of the parking lot under surveillance by camera C5.

Figure 6: Experimental network camera views deployed in area depicted by Figure 5. Each of the six views used in the experimental results are shown. (1) A one-way street leads to (2) the same street, a block away approaches (3) a 4-way intersection right turns from this intersection lead to (4) a side street. Vehicles that move straight through the intersection reach (5) a one-way street. (6) A parking lot can be reached from the road observed by camera four.

Camera color balance, shutter and gain were adjusted to minimize the differences between cameras prior to deployment. Once placed, cameras were stationary throughout the experiments. Each camera was placed to observe a different portion of the street network shown in Figure 5. Figure 6 shows the six different views corresponding to each of the cameras used in the experiment.

Once deployed, the system was run for 374 consecutive minutes during the day. Table 1 reports how many objects were successfully tracked during the period in each view. For each of these objects, an appearance model was constructed on the local computer and transmitted with appropriate timestamps to the server for matching and model update.

TABLE 1: NUMBER OF OBJECTS GENERATED BY EACH CAMERA OVER 374 MINIUTES.

C1	C2	C3	C4	C5	C6
922	806	2,244	1734	1,026	7 4

The number of times objects were seen in two views but not matched by the system based on tight threshold constraints was estimated by hand analysis of the data to be approximately 62%. However, the false positive rate during the experiment was reduced to approximately 2%. The computed transition matrix at the end of the experiment is given by:

$$\begin{bmatrix} 0.0 & 0.95 & 0.02 & 0.0 & 0.0 & 0.01 & 0.02 \\ 0.03 & 0.0 & 0.94 & 0.01 & 0.0 & 0.01 & 0.01 \\ 0.06 & 0.03 & 0.0 & 0.15 & 0.02 & 0.64 & 0.10 \\ 0.01 & 0.02 & 0.35 & 0.03 & 0.17 & 0.03 & 0.39 \\ 0.0 & 0.01 & 0.0 & 0.0 & 0.02 & 0.0 & 0.97 \\ 0.01 & 0.02 & 0.0 & 0.1 & 0.01 & 0.0 & 0.95 \\ 0.64 & 0.02 & 0.01 & 0.23 & 0.04 & 0.06 & - \end{bmatrix}$$

The ability of the matrix to correctly predict the path of new objects was then tested. Cameras captured 10 minutes of traffic data that was analyzed by hand. Twenty vehicles, entering the network at camera C1 were randomly selected and their paths noted by human analysis of the video data. These baseline trajectories were then compared to the predictive power of the network. Given the learned topological model, the probability that an object, currently in camera C_l will transition through sequence $C_l, \ldots, C_{k+l}$ is provided by the composition of the transition probabilities in the learned topological model:

$$p(C_1,\ldots,C_k \mid A) = \left[\prod_{i=1}^{k} A(q_i \mid q_{i-1})\right] A(q_{k+1} \mid q_k) \qquad \text{(Eq. 3)}$$

Where $A(q_i|q_{i-1})$ is the transition probability to state i from state $i\text{-}1$ corresponding to camera i and camera $i\text{-}1$ respectively. For each of the twenty ground truth vehicles, prediction accuracy was computed by measuring the predicted camera position for each object, given its current location k steps into the future. A predicted location is the state that is k steps from the current state and maximizes equation 3. Mean percentage of correct predictions for all twenty vehicles as they moved through the network are shown in Table 2. Mean values for a particular look-ahead value are measured as the average correct prediction rate measured from all states that the object moved through in the network. For example, the accuracy of prediction for a look ahead value of one was measured by predicting the next view a vehicle will be seen in given it is currently in camera C_1, C_2, etc.

TABLE 2: PERCENTAGE OF CORRECTLY PREDICTED OBJECT MOTION FOR LOOK AHEAD VALUES RANGING FROM 1 TO 5

1	2	3	4	5
0.92	0.88	0.58	0.13	0.14

Results indicate that the topological model is capable of accurately predicting object motion from one camera to the next quite accurately. Although not depicted here, prediction accuracies vary based on the state that corresponds to the position of the object. For example, the model was capable of predicting object motion from C_1 to C_2 along the one-way street with greater than 96% accuracy while cameras observing areas with more variability yield less information about likely object paths. This is also the case for look ahead values father than only a few cameras into the future. As look ahead values increase the number of potential paths, even in a simple domain such as the one tested here, grows dramatically and predictions have very little value.

5. CONCLUSION

Topological calibration is a useful goal in geospatial video surveillance scenarios. Acquired models can be used to predict object paths through the surveillance network and analyze traffic flow in complex environments. The model can assist users of the network by presenting likely next views to an operator who is interested in observing subject motion over a wide-area. The principles here can be applied to tracking people and other domains using multiple sensors with little potential for overlap.

Results demonstrate that a global topological model can be acquired and used to predict object motion with greater than 85% accuracy. In future work, we are exploring more sophisticated matching algorithms that are specifically designed to robustly match appearance-based models across widely separated views. In initial work, we are designing radiometric transfer functions that model expected pixel value changes between sensors.

A more accurate initial estimate of both network topology and transition probabilities will decrease time to convergence to a more accurate model. Convergence times vary depending on traffic density, matching robustness, and the configuration of the network. However, a method that allows users to provide an initial transition matrix will certainly speed convergence. Finally, we are extending the method to other domains that include, environmental monitoring of animals at a controlled preserve and indoor surveillance of people for security applications.

ACKNOWLEDGMENT

The author would like to acknowledge the support of Xiong Quanren in deploying the experimental camera network and collecting data. The distributed vision API, *libeasycamera* (available at www.metaverselab.org) developed by Nathan Sanders was used to capture images from the cameras, process, and transmit images as part of this work.

REFERENCES

[1] J. Bilmes, "A Gentle Tutorial on the EM Algorithm and its Application to Parameter Estimation for Gaussian Mixture and Hidden Markov Models". Technical Report, University of Berkeley, ICSI-TR-97-021, 1997.

[2] R.T. Collins, A. Lipton, H. Fujiyoshi, and T. Kanade, "Algorithms for Cooperative Multisensor Surveillance ", In *Proceedings of the IEEE*, October 2001, Vol. 89, No.10, pp 1456-1477.

[3] R. T. Collins, Y. Tsin, J.R. Miller, and A. Lipton. "Using a DEM to Determine Geospatial Object Trajectories." CMU technical report CMU-RITR -98-19, 1998.

[4] T. Darrell, D. Demirdjian, N. Checka, and P. Felzenswalb. "Plan-view trajectory estimation with dense stereo background models." *In*

Proceedings of the International Conference on Computer Vision, Vancouver, BC, July 2001.

[5] S. Dockstader and A. M. Tekalp, "Multiple Camera Fusion for Multi-Object Tracking", *IEEE Workshop on Multi-Object Tracking*, 2001.

[6] A. Elgammal, S. Lim, and L. Davis, "A Reconfigurable and Scalable Multi-Camera Visual Surveillance System", *In Proc. of International Conference on Multimedia and Expo*, 2003.

[7] R. Gross, J. Yang, A. Waibel, "Growing Gaussian Mixture Models for Pose Invariant Face Recognition", *Proceedings of the International Conference on Pattern Recognition (ICPR'00)-Volume 1*, 2000.

[8] Haritaoglu and M. Flickner, "*Detection and tracking of shopping groups in stores*," in Proc. IEEE Conf. on Computer Vision and Pattern Recognition, Kauai, Hawaii, 2001.

[9] M. Harville, G. Gordon, and J. Woodfill. "Foreground segmentation using adaptive mixture models in color and depth", *In Proc. Workshop on Detection and Recognition of Events ion Video*, 2001.

[10] T. Horprasert, D. Harwood, and L. Davis, "A Statistical approach for Real-time Robust Background Subtraction and Shadow Detection", In *Proc. Frame Rate Workshop, in conjunction with ICCV*, 1999.

[11] O. Javed, S. Khan, Z. Rassheed, and M. Shah, "Camera Handoff: Tracking in Multiple Uncalibrated Stationary Cameras", In *Proc. Workshop on Human Motion*, pp. 113-121, 2000.

[12] C. Jaynes, "Multi-View Calibration from Planar Motion for Video Surveillance", *IEEE Workshop on Video Surveillance, in Conjunction with CVPR'99,* Nov. 1999.

[13] C. Jaynes and J. Hou, "Temporal Registration Using a Kalman Filter for Augmented Reality Applications", In *Proc. Vision Interface Conference*, May 2000.

[14] N. Johnson and D. C. Hogg. "Learning the distribution of object trajectories for event recognition." *Image and Vision Computing*, vol. 14, pp. 609-615, 1996.

[15] V. Kettnaker and R. Zabih, "Bayesian multi-camera surveillance", in Proc. of IEEE Conf. On Computer Vision and Pattern Recogntion, Fort Collins, CO, pp. 253-259, 1999.

[16] J. Krumm, S. Harris, B. Meyers, B. Brumitt, M. Hale, and S. Shafer, "Multi-camera multi-person tracking for EasyLiving," In Proc. *IEEE Intl. Workshop on Visual Surveillance*, Dublin, Ireland, 2000, pp. 3--10.

[17] L. Lee, R. Romano, and G. Stein. “Monitoring activities from multiple video streams: Establishing a common coordinate frame”, *IEEE Trans. on Pattern Analysis and Machine Intelligence*, (Special Issue on Video Surveillance and Monitoring), 2000.

[18] J. van Leuven, M.B. van Leeuwen, F.C.A. Groen, “Real-time Vehicle Tracking in Image Sequences”, *In Proc. IEEE Instrumentation and Measurement Technology Conference*, 2001.

[19] H. Pasula, S. Russell, M. Ostland, Y. Ritov, "Tracking Many Objects with Many Sensors" In *International Joint Conference on Artificial Intelligence*, Stockholm 1999

[20] C. Stauffer, W.E.L. Grimson “Adaptive Background Mixture Models for Real-Time Tracking “ *In Proc. IEEE Conference on Computer Vision and Pattern Recognition*, June 23-25, 1999.

[21] C. Stauffer and W. E. Grimson, “Learning Patterns of Activity Using Real-Time Tracking”, *IEEE Trans. On Pattern Analysis and Machine Intelligence*, vol. 22(8), pp. 747-757, 2000.

[22] A. Kropp, N. Master, and S. Teller, “Acquiring and Rendering High-Resolution Spherical Mosaics”, *In Proc. Workshop on Omnidirectional Vision, in Conjunction with CVPR*, 2000.

[23] Wada, T., Tamura, M., and Matsuyama, T., “Cooperative Distributed Object Identification for Wide Area Surveillance Systems”, *Proc. of the Meeting on Image Recognition and Understanding* (MIRU'96), pp.103-108, 1996.

Generation and Application of Virtual Landscape Models for Location-Based services

Norbert Haala and Martin Kada

Institute for Photogrammetry (ifp), University of Stuttgart, Germany Geschwister-Scholl-Strasse 24D, D-70174 Stuttgart
Norbert.Haala@ifp.uni-stuttgart.de

ABSTRACT

The efficient collection and presentation of geospatial data is one key task to be solved in the context of location based services. As an example, virtual landscape models have to be generated and presented to the user in order to realize tasks like personal navigation in complex urban environments. For an efficient model generation, different data sources have to be integrated, whereas an efficient application for location based services usually requires the use of multiple sensor configurations. The work on the generation and application of virtual landscape models described within the paper is motivated by a research project on the development of a generic platform that supports location aware applications with mobile users.

1. INTRODUCTION

The ongoing rapid developments in the field of computer graphics meanwhile allow the use of standard hard- and software components even for challenging tasks like the real-time visualization of complex three-dimensional data. As a result, components for the presentation of structured three-dimensional geodata are integrated in an increasing number of applications. If virtual three-dimensional landscapes and building models – both indoor and outdoor – are visualized three-dimensionally, the access to spatial information can for example be simplified within personal navigation systems. In order to realize these type of applications, 3D landscape models have to be made available as a first step and tools allowing for the efficient presentation of this data have to be provided.

Within the paper, we present our work on the generation and application of virtual landscape models. These algorithms were developed as a part of the Nexus project, which was started at the University of Stuttgart, Germany, with the goal of developing concepts and methods for the support of mobile and location-based applications. Meanwhile, this project has been extended to the interdisciplinary center of excellence "World Models for Mobile Context-Aware Systems", covering issues concerning communication, information management, methods for model representation and sensor data integration

(Stuttgart University 2003). One of the long term goals of this project is the development of concepts and techniques for the realization of comprehensive and detailed world models for mobile context-aware applications. In addition to a representation of stationary and mobile objects of the real world these world models can be augmented by virtual objects, and objects of the real world can be linked to additional information. The result is the so-called "Augmented World Model", which is an aggregated model of the real world and a symbiosis of the real world and digital information spaces. The complexity of these world models ranges from simple geometric models, to street maps and to highly complex three-dimensional models of buildings. In the following section, the data collection for the virtual landscape model, which is used as a basis for our investigations is described. In the second part of the paper, the visualization of this model and data access is discussed.

2. DATA COLLECTION

For our investigations a detailed virtual landscape world model of the city of Stuttgart and the surrounding area of the size 50x50km was made available. The data set includes a 3D city model, a digital terrain model and corresponding aerial images for texture mapping.

2.1 Integration of existing data

Since the development of tools for the efficient collection of 3D city models has been a topic of intense research in recent years, meanwhile a number of algorithms based on 3D measurement from aerial stereo imagery or airborne laser scanner data are available. A good overview on the current state-of-the-art of experimental systems and commercial software packages is for example given in (Baltsavias, Grün, van Gool 2001). Due the availability of these tools a number of cities already provide area covering data sets, which include 3D representations of buildings.

For our test area, a 3D city model was collected on behalf of the City Surveying Office of Stuttgart semi-automatically by photogrammetric stereo measurement from images at 1:10000 scale (Wolf 1999). For data collection, the outline of the buildings from the public Automated Real Estate Map (ALK) was additionally used. Thus, a horizontal accuracy in the centimeter level as well as a large amount of detail could be achieved. The resulting model contains the geometry of 36,000 buildings represented by 1.5 million triangles. In addition to the majority of relatively simple buildings in the suburbs, some prominent historic buildings in the city center are represented in detail by more than 1,000 triangles each. An overview visualization based on the available data is given in Figure 1.

Figure 1: Overview of the Stuttgart city model covering a total of 36,000 building models.

2.2 Texture Mapping

Image texture for visualizations similar to Figure 1 is usually provided from ortho images, which can be collected by airborne or spaceborne sensors. For visualizations from pedestrian viewpoints, like they are required for navigation applications, the visual appearance of buildings has to be improved. For this reason, façade texture was additionally collected for a number of buildings in the historic central area of the city. Whereas ongoing research aims at automating of this process, within the first phase of the project manual mapping was applied for this purpose.

2.2.1 Manual Mapping

This manual mapping of the facades was based on approximately 5,000 terrestrial images collected by a standard digital camera. From these images, which were available for approximately 500 buildings, the façade textures were extracted, rectified and mapped to the corresponding planar segments of the buildings using the GUI depicted in Figure 2.

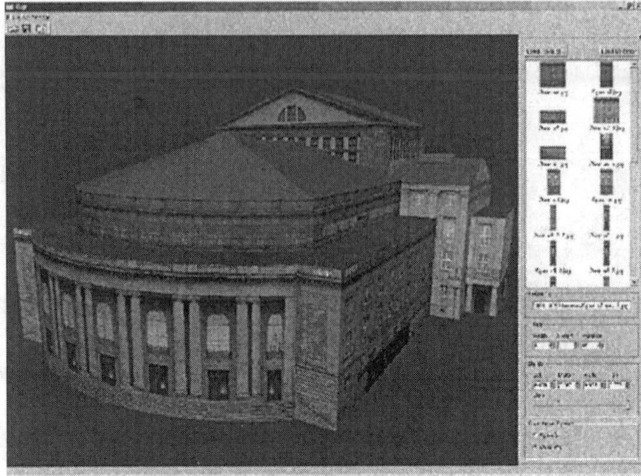

Figure 2: GUI for manual texture mapping of façade imagery.

This GUI allows the user an easy selection of corresponding points at the façade and the respective images. Based on this information the effects of perspective distortion are eliminated by a rectification and the resulting image is then initially snapped to the corresponding part of the building model to be textured. A precise adjustment of the final texture coordinates is then realized by a user controlled affine transformation. Finally, in order to reduce the partly large size of the original images, the texture images are down-sampled to a resolution of approximately 15 cm per pixel at the facades.

Figure 3: Rendered view of textured building models.

A visualization based on the result of manual texture mapping is depicted in Figure 3. In this example additionally random colors were assigned for buildings in the background of the scene, were no real image texture from manual mapping was available.

2.2.2 Panoramic images

One option to provide real image texture at lower quality, but a reduced effort compared to manual mapping is the application of panoramic images. For this purpose we used the high resolution digital panoramic camera TOPEYE, originally developed as a measurement system for photogrammetric purposes (Scheibe et al. 2001). Based on a CCD line, which is mounted on a turntable parallel to the rotation axis, high resolution 360 degree panoramic images can be generated. In order to reach the highest resolution and a large field of view, a CCD line with about 10.000 detector elements is used. The second image dimension is generated by rotating the turntable. Since this CCD is a RGB triplet it allows for the acquisition of true color images.

Figure 4: Image collected by the panoramic camera EYESCAN.

Figure 4 depicts a complete scene collected by the panoramic camera from the top of a building. The enlarged section demonstrates the high resolution, which can be reached by this type of camera. If as in this example, the scene gives a good overview of a larger area, texture mapping is feasible for a number of buildings at least with a limited amount of detail. If the exterior orientation of the panoramic image is available, this can be realized automatically similar to the generation of ortho images. In order to determine the required orientation parameters, corresponding image points can be measured for a limited number of known object points and then used as control points during spatial resection. These control points can for example be provided from the available 3D building models. Alternatively, the exterior orientation can be directly measured as it is described in the following section.

2.2.3 Automated texture mapping from directly georeferenced terrestrial images

If during image collection the position and orientation of the camera is directly measured at a sufficient accuracy, corresponding image coordinates can be calculated for the depicted 3D building model. These correspondences allow for automatic texture mapping without any additional manual effort. For this reason the platform depicted in Figure 5 was used to collect directly georeferenced terrestrial scenes.

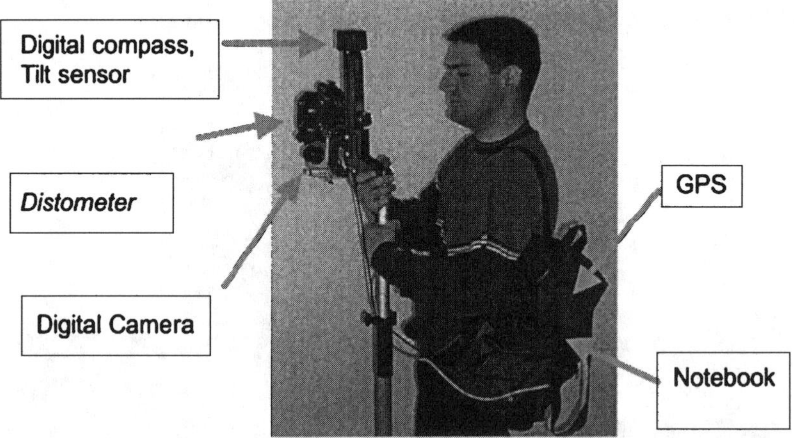

Figure 5: Low cost device for the collection of oriented terrestrial images.

The platform combines a standard resolution digital camera with an extremely wide-angle lens, a GPS receiver, an electronic compass and a tilt sensor. All the devices are connected to a laptop. While the camera and compass/tilt sensor are hand held, the GPS is attached to a backpack. The camera was pre-calibrated to avoid problems due to lens distortions. The GPS receiver is a Garmin LP-25, which can be operated both in normal and differential modes. In our application the ALF service (Accurate Positioning by Low Frequency) was used to receive a correction signal for differential mode processing every three seconds. While the theoretical accuracy of differential GPS as it is used in the prototype is very high, there are a number of practical limitations when this techniques is applied in built-up areas. Shadowing from high buildings can result in poor satellite configurations, and in the worst case the signal is lost completely. Additionally, signal reflections from buildings nearby can give rise to so called multipath effects, which are further reducing the accuracy of GPS measurement. Our experience shows that the system allows for a determination of the exterior orientation of the camera to a precision of 7-10 m in planar coordinates. In our system, the vertical component of the GPS measurement was discarded and substituted by height values from a Digital Terrain Model due to the higher accuracy of that data source. The zenith angle provided by the tilt sensor has an error of approximately 1° – 2°. The applied digital compass is specified to provide the azimuth with a standard deviation of 0.6° to 1.5°. However, compasses are vulnerable to distortion, because especially in build-up areas the Earth's magnetic field can be influenced by cars or electrical installations. These disturbance can reduce the accuracy of digital compasses to approximately 6° (Hoff and Azuma 2000).

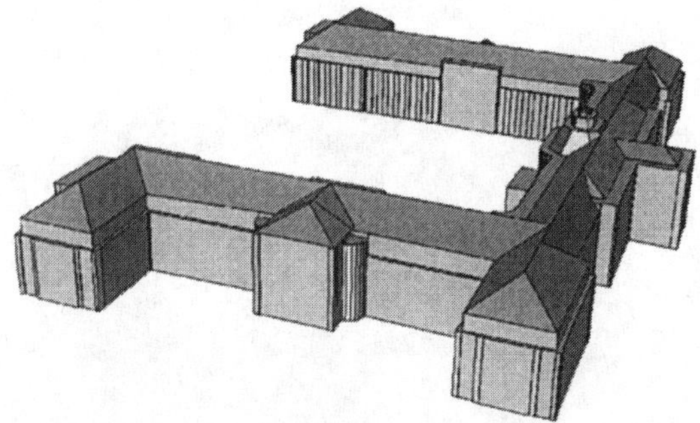

Figure 6: Available 3D building model.

Figure 7: Projected building model from directly measured exterior orientation.

The limited mapping accuracy, which results from the restricted accuracy of the directly measured exterior orientation by our system is demonstrated in Figure 6 and Figure 7. Figure 6 shows a rendered 3D view of the a building model as it is available from the data set already depicted in Figure 1. This model is then overlaid to the image in Figure 7 based on the measured orientation and calibration of the camera. The deviations between model and image are clearly visible. Of course, the quality of direct georeferencing can be improved, if for example inertial sensors are applied. Still, since one of our main goals was the provision of a low-cost system, this was not an option for our application. Alternatively, this coarse model to image mapping was refined by the application of a Generalized Hough Transform (Haala and Böhm 2003). By this approach the visible silhouettes of the depicted buildings are localized automatically in the image.

Figure 8: Building model overlaid to the image based on improved exterior orientation.

The outline of the projected building model, which is used for this purpose, is represented by the yellow polygon in Figure 7. This shape can then be detected based on the Generalized Hough Transform (GHT) no matter whether it is shifted, rotated or optionally even scaled in relation to the respective image. Additionally the GHT allows for a certain tolerance in shape deviation. This is also necessary, since the CAD model of the building provides only a coarse generalization of its actual shape as it is appearing in the image. After the localization of the outline of the building model, check points can be generated automatically based on the 3D coordinates of the visible building and used for the improvement of the exterior orientation by a spatial resection. Based on this information, the mapping between image and model can be refined as depicted in Figure 8. Afterwards, image texture can be extracted automatically for the visible facades.

3. VISUALIZATION AND DATA ACCESS

During personal navigation, which is one of the main tasks within location based services, the visualization of the environment and the generation of a virtual walk through for planning of actual tours are features of great importance. Due to the large amount of geometry and texture data contained in a virtual city model, a brute force rendering approach is not suited even for current high performance 3D graphics accelerators. It is therefore inevitable that we use acceleration techniques like visibility culling, level of detail (LOD) representations and image based rendering in order to speed up the visualization process.

3.1 Impostors

Impostors are an image based rendering technique that allow for a considerable speed up during the visualization of building objects (Schaufler 1995). An impostor replaces a complex object by an image that is projected to a transparent quadrilateral. These images are dynamically generated by rendering the objects for the current point of view. For consecutive, contiguous viewpoints, the impostor images of objects that are located far from the viewer do not change notably with every frame. This allows reuse of impostor images for several frames and therefore speed up the rendering process.

In our work the application of impostors is implemented in Open Scene Graph (OSG), which is a cross-platform C++/OpenGL library for real-time visualization. Depending on a user-defined distance threshold, the building objects as they are provided from the 3D city model are either rendered traditionally or as an impostor image. The recomputation of the impostor image is performed automatically using a pre-defined error criterion. Experimental results on a standard PC equipped with a 2.0 GHz Intel Pentium P4 processor, 512 MB of memory and an NVIDIA GeForce4 Ti4200 graphics accelerator with 128 MB of graphics memory showed a speed up of 350% for our data set.

3.2 Geometric simplification

Whereas impostors provide good results for the visualization of buildings relatively far away from the current point of view, geometric simplification is more advantageous for buildings at closer distance to the virtual observer. Thus, a generalisation process was developed, which automatically generates different levels of details for the respective buildings (Kada 2002). During generalisation, unnecessary details of the buildings are eliminated, whereas features, which are important for the visual impression like regular structures and symmetries, are kept. In our approach the simplification of the polyhedral building models is achieved by combining techniques both from 2D cartographic generalization and computer graphics. During our generalization, symmetries and regularities of the buildings are stringently preserved by integration of a set of surface classification and simplification operations. The initial step of the generalisation algorithm is to build the so-called constrained building model, which represents the regularization constraints between two or more faces of the polyhedral building model. In the following steps the geometry of the constraint building model is then iteratively simplified by detection and removal of features with low significance to the overall appearance of the building. During this feature removal step the constrained building model is applied in order to preserve the represented building regularities and optimise the position of the remaining vertices.

Figure 9: Original building model (with texture).

Figure 10: Simplified building model (with texture).

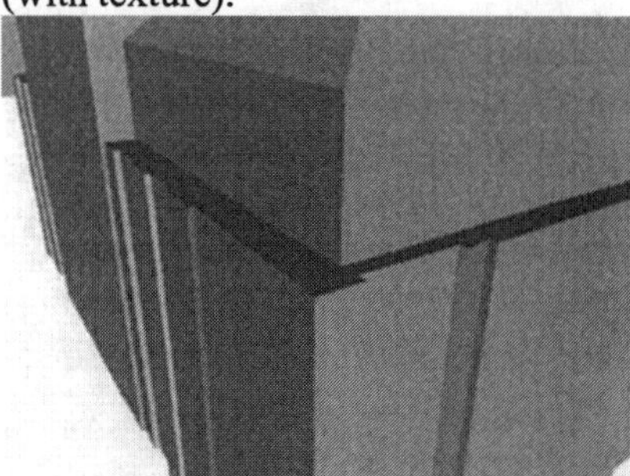

Figure 11: Original building model (without texture).

Figure 12: Simplified building model (without texture).

The result of our algorithm is demonstrated for a part of a building in Figure 9 to Figure 12. Figure 9 and Figure 11 show a part of the original model as it was captured from stereo imagery and an existing outline from the public Automated Real Estate Map (ALK), respectively. Figure 12 shows the result of the generalisation process. It is clearly visible, that parallelism and rectangularity have been preserved for the remaining faces. Especially, if the model is textured again as it is depicted in Figure 10, this amount of detail is sufficient for realistic visualization even at close distances.

3.3 Information access by oriented images

In addition to a realistic visualization, location aware services require the provision of tools allowing for an intuitive access to object related information. This can also be realized based on the georeferenced images as they are collected by our low-cost system depicted in Figure 5. As it is demonstrated in Figure 8, the available 3D model is co-registrated to a real image of the environment as it is perceived by the user. Thus, access to localized information is feasible by pointing to the respective regions of interest within the image.

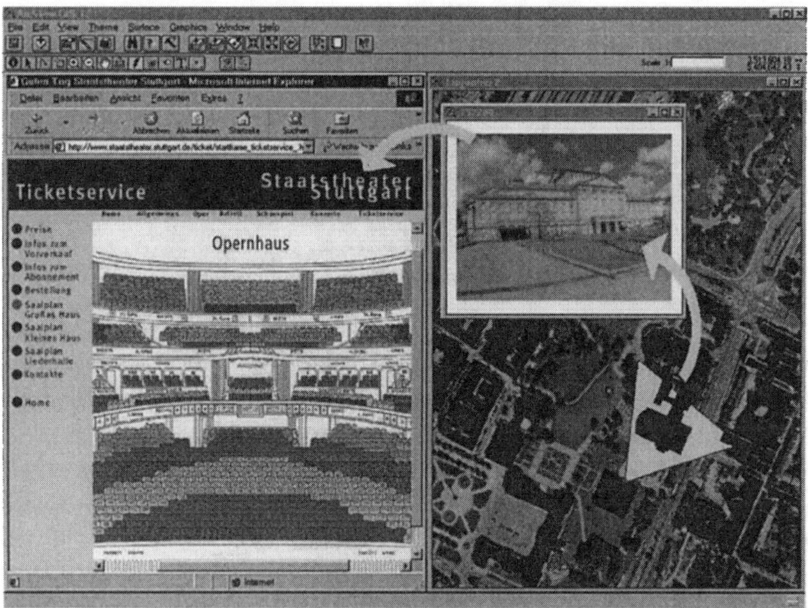

Figure 13: Exemplary application demonstrating the access to object related information.

An exemplary application of our current prototype is depicted in Figure 13 (Haala 2001). Since the position and orientation of the user is available, the visible building is selected from the data base and corresponding object related information as it is for example provided by a website is presented by the graphical user interface. These websites then give access to services like ticket sales if for example a theatre is visible. Additionally, the user's location and the selected building can be projected to an ortho image or a map. For demonstration of the telepointing functionality, this application is realized within a standard GIS software package.

The overlay of computer graphics representing object related information to the user's current field of view is a standard feature within augmented reality (AR) applications. As an example, a mapping between a building model and a real image of the environment as it is discussed within this paper can for example also be applied to present virtual reconstructions of devastated buildings to visitors of historical sites (Vlahakis et al 2002). Since an emerging spread of these types of concepts can be expected, they will have a considerable impact on future applications of virtual landscape models for location-based services.

4. REFERENCES

Baltsavias, E., Grün, E. and van Gool, L., 2001. *Automatic Extraction of Man-Made Objects From Aerial and Space Images (III).*

Haala, N., 2001. Automated Image Orientation in a Location Aware Environment. *Photogrammetric Week 2001*, pp. 255-262.

Haala, N. and Böhm, J., 2003. A Multi-Sensor System for Positioning in Urban Environments. *ISPRS J. of Photogrammetry and Remote Sensing* **58**(1-2), pp. 31-42.

Hoff, B. and Azuma, R., 2000. Autocalibration of an Electronic Compass in an Outdoor Augmented Reality System. *Proceedings of International Symposium on Augmented Reality*, pp.159-164.

Kada, M., 2002. Automatic Generalisation of 3D Building Models. *IAPRS* Vol. 34, Part 4, on CD.

Schaufler, G., 1995. Dynamically Generated Impostors. *GI Workshop "Modeling - Virtual Worlds - Distributed Graphics"*, pp.129-135.

Scheibe, K., Korsitzky, H., Reulke, R., Scheele, M. and Solbrig, M., 2001. EYESCAN - A High Resolution Digital Panoramic Camera. *RobVis 2001*, pp.77-83.

Stuttgart University, 2003. Nexus: World Models for Mobile Context-Based Systems. http://www.nexus.uni-stuttgart.de/.

Vlahakis, V., Ioannidis, N., Karigiannis, J., Tsotros, M., Gounaris, M., Stricker, D., Gleue, T., Daehne, P. and Almeida, L., 2002. Archeoguide: An Augmented Reality Guide for Archeological Sites. *Computer Graphics and Applications* **22**(5), pp. 52-60.

Wolf, M., 1999. Photogrammetric Data Capture and Calculation for 3D City Models. *Photogrammetric Week '99*, pp. 305-312

A Low-Cost System for Creating 3D Terrain Models from Digital Images

Howard Schultz

Aerial Vision Inc.
Amherst, MA 01002
schultz@aerialvision.com

ABSTRACT

Recent advances in digital image sensors and computer vision techniques have the potential to significantly improve the ability to monitor and study a wide variety of environments, including the relationship between forest conservation and global climate change, the impact of suburban sprawl on drinking water quality, and fire risk assessment over large tracts of public and private land. However, the full potential of these technologies cannot be realized until robust, automatic end-to-end systems are made available that enable the end user to generate meaningful results from these new data sources and analysis tools. Our goal is to improve the way in which Geographic Information System (GIS) databases are created, updated and utilized. We are building systems that enable organizations to rapidly and inexpensively generate, update, analyze and visualize high-resolution 3D digital terrain models from digital images collected by a portable camera system mounted on a single engine aircraft. These techniques will expand the scope of GIS applications, especially in areas where low cost and short turnaround time are critical. This paper discusses the general problem of reconstructing 3D terrain models for small format digital images, and presents a system that integrates an instrument package (constructed from commercial off-the-shelf components) with a suite of analysis and terrain modeling algorithms. Preliminary results are presented that demonstrate the feasibility of the system.

1. INTRODUCTION

Understanding the impact of society on the natural world has become a central theme in scientific research and government policy formation. U.S. government agencies and international organizations are increasing the amount of resources allocated to studying a wide range of environmental and ecological topics, including carbon cycle modeling, global warming, bio-diversity and resource management. To better understand the impact of these programs high-resolution terrain models and attribute maps over ecologically sensitive areas need to be quickly and economically generated. The primary goal of our research and development program is to enhance the ability of ecology and resource management programs to acquire, analyze and distribute high-resolution maps of important environmental attributes.

This paper describes a remote sensing system under development by Aerial Vision Inc. (AVI) for collecting high quality, high-resolution digital images and metadata, and creating geographically registered high-resolution 3D terrain models and attribute maps. The goal is to provide a user friendly end-to-end system that will enable scientists and engineers with a minimum of specialized training in photogrammetry and computer vision to quickly and inexpensively generate high quality GIS attribute layers. The system is based on several years of research in computer science and resource management at the University of Massachusetts, Amherst [Schultz 99b, 99c, 02b; Slaymaker 99].

2. GENERAL CONSIDERATIONS

The process of creating 3D terrain models and attribute layers involves a wide range of technologies from camera calibration to softcopy photogrammetry. This paper examines each step in the process with the goal of determining a proper mix of off-the-shelf components and new technologies for building a practical system.

In theory, a 3D terrain model can be reconstructed from a collection of images if the camera(s) are calibrated so that the orientation of the image rays are known relative to the camera coordinate system, and the position and orientation (pose) of the camera, for each exposure, is known relative to a fixed world coordinate system. The intrinsic camera parameters are determined in the laboratory. The camera pose, however, must be determined from metadata collected during data acquisition.

Our approach uses an optimal filtering technique to produce accurate estimates of camera pose from reliable but noisy estimates from low-cost instruments. Softcopy photogrammetry algorithms further reduces the measurement error to a level that permits the generation of geographically registered 3D terrain models.

Camera Calibration. Camera calibration is an essential element that is often overlooked in the construction of low cost systems. The purpose of geometric camera calibration is to build a mathematical model that defines the orientation of the rays that emanate from each pixel. Traditional calibration procedures, based on this philosophy, categorize the camera and lens separately. These methods work well provided that great care is taken in the manufacture and handling of the camera and lenses. For low-cost cameras, however, it is not possible to consider the components separately. Instead, the camera and lens must be treated as one unit.

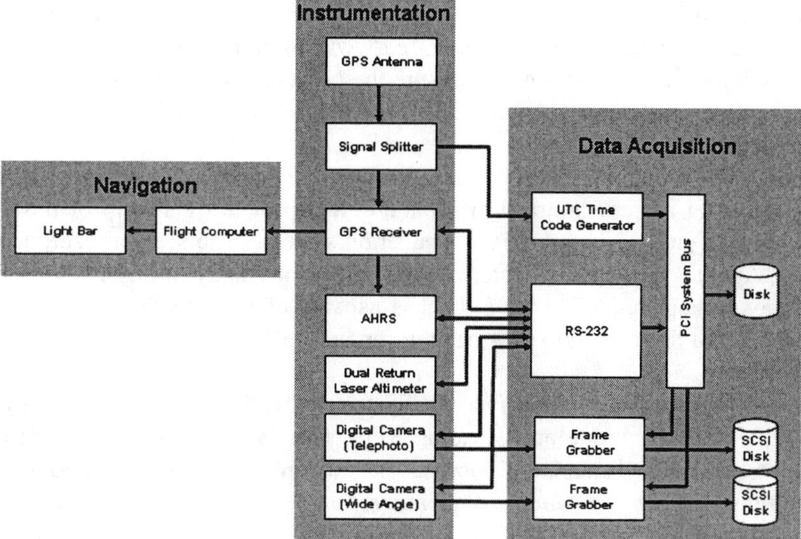

Figure 1: A data flow diagram of the navigation, instrumentation and data acquisition systems. The signal received by the GPS antenna is split between the GPS receiver and the UTC time code generator. The GPS receiver, AHRS and dual return laser altimeter download data to the system through an RS-232 interface. The digital cameras download data through a high-speed camera link port on the frame grabbers, which write the data directly to disk through an onboard SCSI controller. The GPS receiver, AHRS, laser altimeter, and digital cameras receive setup and control function commands from the system through the RS-232 communications ports. The pilot is guided by a light bar and flight computer.

The intrinsic camera parameters (focal length, image center and lens distortion) are determined using the calibration procedures developed at the University of Massachusetts, Amherst [Kovalenko 02].

3. INSTRUMENTATION AND DATA ACQUISITION

A system for collecting multi-spectral images and navigation metadata is currently under development. The system is built from commercial-off-the-shelf (COTS) parts and is designed to be maintained and operated with a minimum of specialized training. To ensure optimal flexibility, the system is designed so that it can be mounted to a Cessna 172/182 aircraft, which are readily available for rent at most general aviation airports in most countries. The basic system components are shown in Figure 1 and described below.

Laser Altimeter. The system incorporates a profiling laser altimeter, manufactured by Laser Atlanta [http://www.laseratlanta.com]. The primary function of the altimeter is to measure the instantaneous height of the aircraft above the terrain at approximately 240 Hz.

Navigation System. During data collection, the navigation system guides the pilot along a predetermined grid pattern, and provides a stream of position estimates (at 5 Hz) to the data acquisition system. The system incorporates an Ag-Vav navigation system, which was originally designed for crop spraying applications. The system is capable of locating the instantaneous position of the aircraft to sub-meter accuracy at 5Hz.

Attitude and Heading Reference System (AHRS). A Watson Industries AHRS E304 Attitude and Heading Reference System [http://www.watson-gyro.com/products/ahrs.html] is used to determine the camera attitude and heading (where the camera is pointing) to an accuracy of 0.01°, and 0.1° relative to the vertical plane.

Data Acquisition System. The data acquisition system platform is a standard 3 GH Pentium workstation. The laser altimeter, ARHS, and GPS receiver generate ASCII data streams at approximately 250 Hz, 17 Hz and 5 Hz, respectively. The data rate from these instruments is slow enough to be read by a standard RS-232 port. The digital cameras, on the other hand, generate data at a significantly higher rate of approximately 80 MB/sec and use a standard high speed camera link interface. Because each instrument operates asynchronously and transmits data at a unique rate, the data from each instrument must be time tagged before it is written to disk.

4. DATA PROCESSING

So far the discussion has focused on procedures for acquiring images and metadata. In the remaining sections, the discussion turns to the process of converting the raw data to high-resolution 3D terrain models.

Pre-processing. The pre-processing procedure provides a simple automatic means for interfacing the raw data to a commercial GIS/softcopy software package. After the data disks are returned to the laboratory, the data are processed by a suite of pre-processing programs that automatically clean and format the raw data.

Camera Pose Recovery. Camera pose is estimated in a two step process. First, the data are unpacked, checked for validity and converted to engineering units, the GPS, AHRS, laser altimeter, and image data are assembled in time coded tabular form, where each record contains the derived observation and the time of the observation.

Next, the independent noisy observations, an estimate of the camera pose for each exposure is derived using a Kalman-Bucy filter [Kalman 61, Gelb 74], which is often used in navigation systems to assimilate position and orientation information from a global positioning system (GPS) and inertial measure unit (IMU) [Lin 1991, Cook 94, Seclel 64]. In our system, photogrammetric analysis of the motion imagery provides additional observations of the aircraft motion [Sim 96].

The camera pose estimates are used to initialize the block bundle adjustment procedure of the softcopy photogrammetry routine. It is not necessary to measure the precise camera pose during flight. Instead, the navigation instruments must simply provide sufficiently accurate estimates of camera pose to initialize the softcopy photogrammetry system.

DEM Extraction. Seamless feature maps are created from a collection of small format digital images by projecting each image onto a digital elevation map (DEM). This process requires precise knowledge of the camera pose at each exposure and the shape of the terrain [Maune 01]. The DEM may be extracted from a database, such as the ones provided by the USGS, or generated from the recorded images using softcopy photogrammetry techniques. For low-resolution projects, an existing DEM may be acceptable. For high-resolution projects requiring sub-meter registration or a dense array of posting, however, the DEM must be extracted from the recorded images and metadata [Rodriguez 90].

We will implement the Terrest methodology [Schultz 94, 95, 02a] to recover high-resolution topography from a sequence of spatially overlapping digital images. Terrest has the capability of fusing the overlapping information captured by a sequence of images into a single composite DEM. The algorithm uses the overlapping information to estimate the optimal elevation, geospatial uncertainty and a reliability figure of merit for a dense array of points [Leclerc 98a, b, Schultz 99b, 02a].

A high resolution DEM and a corresponding orthoimage generated from the images are shown in Figure 2. The ground sampling distance for the DEM is 10 cm. The ortho-image is a false color image with the near infrared, red and green bands encoded in the red, green and blue color channels. The scene covers an area of approximately 100 by 150 meters. Note that

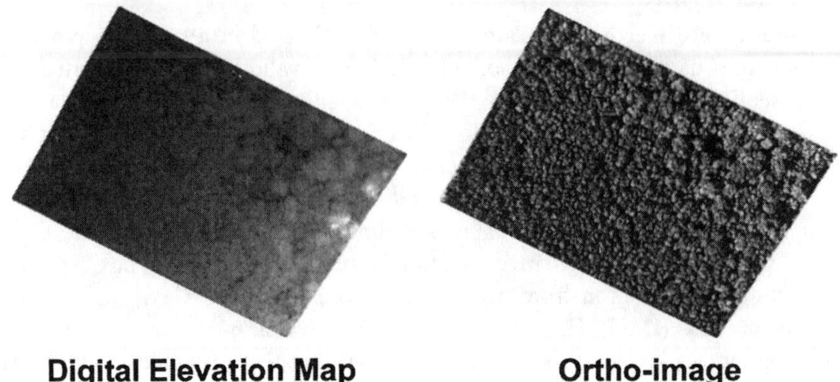

Figure 2: A high resolution DEM and a corresponding ortho-image generated from two overlapping multi-spectral digital images. The ground sampling distance for the DEM is 10cm.

individual trees (and tree gaps) are clearly visible, and that the terrain slopes downhill from right to left.

5. CONCLUSION

Preliminary results have demonstrated the feasibility of building a low-cost, portable aerial imaging and analysis system capable of generating high-resolution 3D terrain models in the form of GIS layers. The instrumentation, data acquisition and navigation system are built from commercial off-the-shelf parts. We have demonstrated the feasibility of generating accurate estimates of camera pose from inexpensive navigation instruments by integrating optimal filtering and standard softcopy photogrammetry techniques.

ACKNOWLEDGMENTS

The work is supported by a grant (DMI-0232361) from the National Science Foundation.

REFERENCES

Blazquez, C.H. 1989. Computer-Based Image Analysis and Tree Counting with Aerial Color Infrared Photography. *Journal of Imaging Technology,* Vol. 15(4) pp. 163-168.

Chester C.S., T. Charles, and W.H. Soren, (Editors), 1980. *Manual of Photogrammetry*, Fourth edition. American Society for Photogrammetry and Remote Sensing, Falls Church, VA.

Cook M. and M. Rycroft, 1994. *Aerospace Vehicle Dynamics and Control*, Clarendon Press, Oxford.

Gelb, A., 1974. *Applied Optimal Estimation*, M.I.T. Press, Cambridge, MA.

Kalman R. and R. Bucy, 1961. New Results in Filtering and Prediction Theory, *J. Basic Eng.*, Ser. D,Vol. 83,pp. 95-108.

Kovalenko, S., 2002. Camera Calibration for Generating Mosaics, M.S. Project, University of Massachusetts, Amherst, Computer Science Department. http://vis-www.cs.umass.edu/~kovalenk/thesis/ms.html.

Leclerc, Y.G., Q.T. Luong, and P. Fua, 1998. A Framework for Detecting Changes in Terrain, *IEEE Trans. Pattern Analysis and Machine Intel.*, Vol. 20(11), pp. 1143-1160.

Leclerc, Y.G., Q.T. Luong et al., 1998. Self-consistency: A Novel Approach to Characterizing the Accuracy and Reliability of Point Correspondence Algorithms, *DARPA Image Understanding Workshop*, Morgan Kauffman.

Lin, C-F. 1991. *Modern Navigation, Guidance, and Control Processing*, Prentice-Hall, Englewood Cliffs, NJ.

Maune, D. (Edtor), 2001. *Digital Elevation Model Technologies and Applications: The DEM Users Manual*, American Society for Photogrammetry and Remote Sensing.

Rodriguez, J.J. and J.K. Aggarwal, 1990. Matching Aerial Images to 3-D Terrain Maps, *IEEE Trans on Pattern Analysis and Machine Intel.* Vol. 12(12).

Schultz, H., A.R. Hanson, E.M. Riseman, F.R. Stolle, D. Woo, and Z. Zhu, 2002. A Self-consistency Technique for Fusing 3D Information. Invited talk at the *IEEE 5th Int. Conference on Information Fusion*, Annapolis, MD.

Schultz, H., A.R. Hanson, E.M. Riseman, and F.R. Stolle, 2002. Rapid Updates of GIS Databases from Digital Images. *National Conference for Digital Government Research*, Los Angeles CA.

Schultz, H., E.M. Riseman, F.R. Stolle, and D.-M. Woo, 1999. Error Detection and DEM Fusion Using Self-Consistency, *7th IEEE Int. Conference on Computer Vision*, Kerkyra, Greece, Vol. 2, pp. 1174-1181.

Schultz, H., A. Hanson, C. Holmes, E. Riseman, D. Slaymaker, and F. Stolle, 1999. Integrating Small Format Aerial Photography, Videography, and a Laser Profiler for Environmental Monitoring, *ISPRS WG III/1 Workshop on Integrated Sensor Calibration and Orientation*, Portland, ME..

Schultz, H., D. Slaymaker, A. Hanson, E. Riseman, C. Holmes, M. Powell, and M. Delaney, 1999. Cost-Effective Determination of Biomass from Aerial Images, *International Workshop on Integrated Spatial Data*, Portland, ME.

Schultz, H., 1995. Terrain Reconstruction from Widely Separated Images, *Proc. SPIE*, Vol. 2486, pp. 113-123, Orlando, FL.

Schultz, H., 1994. Terrain Reconstruction from Oblique Views, *Proc. DARPA Image Understanding Workshop*, Monterey, CA, pp. 1001-1008.

Seckel, E., 1964. *Stability and Control of Airplanes and Helicopters*, Academic Press, NY.

Sim, D.-G., S.-Y. Jeong, R.-H Park, R.-C. Kim, S.U. Lee, and I.C. Kim, 1996. Navigation Parameter Estimation from Sequential Aerial Images, *Int. Conference on Image Processing*, Vol. 1, pp. 629-632.

Slaymaker, D., H. Schultz, A. Hanson, E. Riseman, C. Holmes, M. Powell, and M. Delaney, 1999. Calculating Forest Biomass With Small Format Aerial Photography, Videography and a Profiling Laser, *ASPRS Proc. of the 17th Biennial Workshop on Color Photography and Videography in Resource Assessment*, Reno, NV.

Computer Networks and Sensor Networks

Location-Aware Routing for Data Aggregation in Sensor Networks[1]

Jonathan Beaver, Mohamed A. Sharaf,
Alexandros Labrinidis, Panos K. Chrysanthis

Advanced Data Management Technologies Laboratory
Department of Computer Science
University of Pittsburgh
Pittsburgh, PA 15260, USA
{*beaver, msharaf, labrinid, panos*}*@cs.pitt.edu*

ABSTRACT

In-network aggregation has been proposed as one method for reducing energy consumption in networked sensors. In this paper, we explore the idea of *influencing the construction of the routing trees for sensor networks* with the goal of reducing the size of transmitted data for networks with in-network aggregation involving Group By queries. Toward this, we propose a *group-aware network configuration* method and present two algorithms, that "cluster" along the same path sensor nodes which belong to the same group. We evaluate our proposed scheme experimentally, in the context of existing in-network aggregation schemes, with respect to energy consumption and quality of data. Overall, our routing tree construction scheme provides energy savings over existing network configuration schemes and improves quality of data in systems with imperfect quality of data such as TiNA.

1 INTRODUCTION

From monitoring endangered species [7, 12], to monitoring structural integrity of bridges [8], to patrolling borders, sensor networks today offer an unprecedented level of interaction with the physical environment. Within a few years, miniaturized, networked sensors have the potential to be embedded in all consumer devices, in all vehicles, or as part of continuous environmental monitoring.

Sensor nodes, such as the Berkeley MICA Mote [4] which gathers data such as light and temperature, are getting smaller, cheaper, and able to perform more complex operations, including having mini operating systems embedded in the sensor [5]. While these advances are improving the capabilities of sensor nodes, there are still many crucial

[1]Supported in part by the National Science Foundation award ANI-0325353.

problems with deploying sensor networks. Limited storage, limited network bandwidth, poor inter-node communication, limited computational ability, and limited power still persist.

One way of alleviating the problem of limited power is by employing in-network query processing instead of query processing at the base station. For example, assume a sensor network that is used to monitor the average temperature in a building. One way to implement this is to have each sensor send its temperature reading up the network to the base station, with intermediate nodes responsible for just routing packets. Another way, with in-network query processing (or aggregation), would be for each node to incorporate its own reading with the average computed so far by its children. In this way, only one packet needs to be sent per node and each intermediate node computes the new average temperature before sending information further up the network.

As the example shows, with in-network aggregation some of the computational work of the aggregation is performed within the sensor node before it sends the results out to the network. The reason why in-network aggregation reduces power consumption is that sensor power usage is dominated by transmission costs, as has been shown in [3, 6]. Therefore, being able to transmit less data (the result of the aggregation over having to forward all the packets) results in reduced energy consumption at the sensor nodes.

In this work we explore the idea of *influencing the construction of the routing trees for sensor networks* with the goal of reducing the size of transmitted data, especially with in-network aggregation. More specifically, in addition to traditional link-strength criteria, the idea is to consider the semantics of the query and the properties/attributes of the sensor nodes when configuring the sensor network and in particular building the routing tree for the aggregation. Based on this idea, we propose a *group-aware network configuration* method and developed two algorithms, called *GaNC* and *GaNCi*, that "cluster" along the same path sensor nodes that belong to the same group. The intuition of this approach is that messages along such paths will contain less groups and hence incur less energy cost in transmitting them.

We have experimentally evaluated our proposed group-aware network configuration algorithms using simulation. We have investigated the improvement in energy for group-aware network configuration for the sensor network implementations of TAG and Cougar, which are two representative schemes for in-network aggregation. We have further considered our algorithms in conjunction with a new energy efficient scheme for in-network aggregation called TiNA (Temporal coherency-aware in-Network Aggregation). Our results show that by using group-aware network configuration we have savings in energy of up to 33% over the strongest link method and in the case of TiNA, the proposed method can help improve the quality of data it provides while further increasing energy savings.

The rest of this paper is organized as follows. Section 2 provides an overview of in-network aggregation and the TiNA scheme. Additional background in sensor network routing tree configuration is provided in Section 3. The proposed network configuration algorithms, *GaNC* and *GaNCi*, are presented in Section 4. Section 5 describes our simulation testbed, and then in Section 6 we show our experiments and results. We present related work in Section 7. We conclude in Section 8.

2 BACKGROUND

In this paper we propose network configuration and routing techniques to further save energy in sensor networks. Before presenting the proposed algorithms, we give a brief overview of current in-network aggregation schemes; our proposed techniques work in conjunction with all such schemes.

2.1 In-Network Aggregation

Directed diffusion [2, 6] is the prevailing data dissemination paradigm for sensor networks. In directed diffusion data generated by a sensor node is named using attribute-value pairs. A node requests data by sending interests for named data. Data matching the interest is then drawn towards the requesting node. Since data is self-identifying, this enables activation of application-specific caching, aggregation, and collaborative signal processing inside the network, which is collectively called *in-network processing*. Ad-hoc routing protocols (e.g., AODV[13], Information-directed Routing[9]) can be used for request and data dissemination in sensor networks. These protocols, however, are end-to-end and will not allow for in-network processing. On the contrary, in directed diffusion each sensor node is both a message source and a message sink at the same time. This enables a sensor to seize a data packet that it is forwarding on behalf of another node, do in-network processing on this packet, if applicable, and forward the newly generated packet up the path to the requesting node.

The work on *Cougar* [1, 19] and *TinyDB* [10, 11] introduced the directed diffusion concepts in the database arena. Cougar abstracted the data generated by the sensor network as an append-only relational table. In this abstraction, an attribute in this table is either information about the sensor node (e.g., id, location) or data generated by this node (e.g., temperature, light). Cougar and TinyDB emphasize the savings provided by using in-network aggregation, which is one type of in-network processing. Sensor applications are often interested in summarized and consolidated data that are produced by aggregated queries rather than detailed data.

2.2 Communication in Sensor Networks

Communication in a sensor network can be viewed as a tree, with the root being the base station. Synchronizing the transmission between nodes on a single path to the root is crucial for efficient in-network aggregation. A sensor (parent) needs to wait until it receives data from all nodes routing through it (children) before reporting its own reading. This delay is needed so that the parent node p can combine the partial aggregates reported by its children with its own reading and then send one message representing the partial aggregation of values sampled at the subtree rooted at p. The problem of deciding how long to wait (i.e., synchronize the sending and receiving of messages) is treated differently in Cougar and TAG.

Synchronization in TAG is accomplished by making a parent node wait for a certain time interval before reporting its own reading. This interval, called a *communication slot*, is based on subdivisions of the query period, which is referred to as an *epoch*. During a given communication slot, there will be one level of the tree sending and one level listening. In the following slot, those that were sending will go into *doze* or *sleep* mode until the next epoch, while the nodes that were receiving will now be transmitting. The cycle continues until all levels have sent their readings to their parents. When a parent receives the information, it aggregates the information of all children along with its own readings before sending the aggregate further up the tree. This synchronization scheme provides a query result every epoch duration.

Synchronization in Cougar is motivated by the fact that for a long running query, the communication pattern between two sensors is consistent over short periods of time. Hence, in a certain round, if node p receives data from a node c, then it will realize it is the parent of that node c. Node p will add c to its *waiting list* and predict to hear from it in subsequent rounds. In the following rounds, node p will not report its reading until it hears from all the nodes on its waiting list. However, one case where this prediction fails is when the reading gathered by node c does not satisfy a certain selection predicate and hence needs to be discarded. In this case, under the Cougar protocol, node c will send a *notification packet* to prevent node p from waiting on c indefinitely.

2.3 Temporal Coherency-Aware In-Network Aggregation

TiNA (short for **T**emporal coherency-aware **in**-**N**etwork **A**ggregation) is built as a layer that operates on top of in-network aggregation systems in order to minimize energy consumption throughout the entire sensor network [16]. The current implementation of TiNA has been designed to work with both TAG and Cougar.

TiNA selectively decides what information to forward up the routing tree by applying a hierarchy of filters along each path of the network. The selectivity of TiNA is based on a user specified *TOLERANCE (tct)*. The *tct* value acts as an output filter at the readings level, suppressing readings within the range specified by *tct*. For example, if the user specifies $tct = 10\%$, the sensor network will only report sensor readings that differ from the previously reported readings by more than 10%. Values for *tct* range from 0, which indicates to report readings if any change occurs, to any positive number. This *tct* is the maximum change that can occur to the overall quality of data in the system using TiNA.

A TiNA sensor node must keep additional information in order to utilize the temporal coherency tolerance. The information kept at a certain sensor depends on its position in the routing tree (i.e., a leaf or an internal node). Leaf nodes keep only the *last reported reading* which is defined as the last reading successfully sent by a sensor to its parent. Internal nodes, in addition to the last reported reading for that node, keep the last reported data it received from each child. This data can either be a simple reading reported by a leaf node or a partial result reported by an internal node. Having the last operation repeated at every parent node along all the network paths provides a hierarchy of filters on every path. Setting the *tct* to zero for the hierarchical filtering at intermediate nodes ensures that partial aggregates, and eventually final aggregates, are always within the user-specified tct.

The hierarchy of filters TiNA provides is important for the incremental processing of aggregate queries as it captures cases of temporal correlation that cannot be captured at the readings level by individual sensors. For example, consider the aggregation function SUM; readings from different sensors might change from one round to another, however, it is possible that the overall sum stays the same. This can only be detected at a parent node which intercepts the stream of readings generated by these sensors and acts as an intermediate *centralized* stream processor. Note that this intermediate stream processing can provide a completely empty partial result or a partial result that is missing few aggregate groups when compared to the old partial result. In both cases, this node relies on the fact that its parent stored its last reported data and it will use it to supply the missing groups.

3 ENERGY EFFICIENT DATA ROUTING IN SENSOR NETWORKS

In this work, we assume a sensor grid environment in which the transmission range of each sensor node is one hop (i.e., all neighboring nodes are of equal distance and consume the same transmission energy). This is done to simplify the presentation and to streamline the evaluation of our proposed method. However, our proposed method is directly applicable to the general case (of non-uniform sensor network configurations) as well.

The ability to route data from the various nodes of the sensor network towards a central sink point (i.e., the base station) is fundamental to the operation of sensor networks. To support routing of data, the sensor network is configured into a routing tree, where each node (child) selects a *gradient* [2] or *parent* [10] to propagate its own readings.

The sensor network constructs the routing tree along with the propagation of the query. We assume that a new query in our model originates at the base station which forwards it to the nearest sensor node. This sensor node will then be in charge of disseminating the query down to all the sensor nodes in the network and to gather the results back from all the sensor nodes.

Traditional network configuration methods rely on link strength to construct the routing tree [18]. A child will pick the parent with the highest link strength, since this would usually correspond to shorter distance and thus less energy for transmitting data to the parent.

First-Heard-From Network Configuration The First-Heard-From (FHF) Network Configuration method is a simple way for children to choose parents and thus establish the routing tree. This method is derived from the link strength approach, when the sensor network follows a grid model and the transmission range of sensor nodes is one hop.

The basic idea behind the FHF network configuration algorithm is as follows. Starting from the root node, nodes transmit the new query. Children nodes will select as their parent the first node they hear from and continue the process by further propagating the new query to all neighboring nodes. The process terminates when all nodes have been "connected" via the routing tree.

The FHF method is formally described as follows:

1. The root sensor prepares a query message which includes the query specification. The root sensor also sets the (L_s) value in the message to its level value (i.e., L_{root} which is 0 initially). It then broadcasts this query message to the neighboring sensors.

2. Initially, all sensor nodes have level values set to ∞. A sensor i that

receives a query message and has its level value currently equal to ∞ will set its level to the level of the node it heard from, plus one. That is, $L_i = L_s + 1$.

3. Sensor i will also set its parent value P_i to Id_s. It then will set Id_s and L_s in the query message to its own Id_i and L_i respectively and broadcast the query message to its neighbors.

4. Steps 2 and 3 are repeated until every node i in the network receives a copy of the query message and is assigned a level L_i and a parent P_i.

This routing scheme is simple yet highly effective. It creates a path whereby child nodes can propagate readings up to the root. It also creates a way in which a query message from the root can be received by all nodes in the network. In addition, each node has been assigned a level which is needed for synchronization methods such as the epoch scheme in TAG.

The main weakness of this method is that it creates the network in a random way (only based on network proximity). The children assign parents based on whichever node happened to broadcast the routing message first. This method fails to consider the semantics of the query or the properties/attributes of the sensor nodes and hence it cannot take any opportunities for energy savings. In the next section we present our proposal for an improved network configuration method that alleviates these problems and saves energy.

4 GROUP-AWARE NETWORK CONFIGURATION

In order to have a network configuration method that considers the semantics of the query and the properties of the sensor nodes, we look closely at how in-network aggregation works. In-network aggregation will depend on the query *attributes* and the *aggregation function.* On the one hand, the list of attributes in the Group-By clause subdivides the query result into a set of groups. The number of these groups is equal to the number of combinations of distinct values for the list of attributes. Two readings from two different sensor nodes are only aggregated together if they belong to the same group. On the other hand, the aggregation function determines the structure of the partial aggregate and the partial aggregation process.

For example, consider the case where the aggregate function is SUM. In this case, the partial aggregate generated by a routing sensor node is simply the sum of all readings that are forwarded through this sensor node. However, if the aggregate function is AVERAGE, then each routing sensor node will generate a partial aggregate that consists of the sum of the readings and their count. Eventually, the root sensor node will use the sum and count to compute the average value for each group before forwarding it to the base station for further processing and dissemination.

Because aggregation combines all the readings for a particular group into one aggregate reading, creating a routing tree that keeps members of the same group within the same path in the routing tree should help decrease the energy used. The reason is simple: by *"clustering" along the same path* nodes that belong to the same group, the messages sent from these nodes will contain less groups (i.e., be shorter, thus reducing communication costs).

4.1 Example of Group-Aware Network Configuration

To better illustrate the basic motivation, benefit, and reasoning behind group-aware network configuration, consider the example shown in Figure 1. In this figure, nodes 2, 4, and 6 (the shaded ones) belong to one group, whereas nodes 1, 3, 5, and 7 belong to a different group. Under the standard FHF network configuration (Figure 1a), nodes 4 and 5 could pick 2 as their parent, whereas nodes 6 and 7 could pick 3 as their parent. Using in-network aggregation, the message sizes from nodes 2 and 3 to the root of the network will both be 2 tuples (i.e., contain partial aggregates from two groups). On the other hand, if we were able to cluster along the same path nodes that belong to the same group (Figure 1b) we would reduce the size of messages from nodes 2 and 3 in half: each message will only contain the partial aggregate from a single group (1 tuple). Next, we present the proposed algorithm, which achieves such clustering.

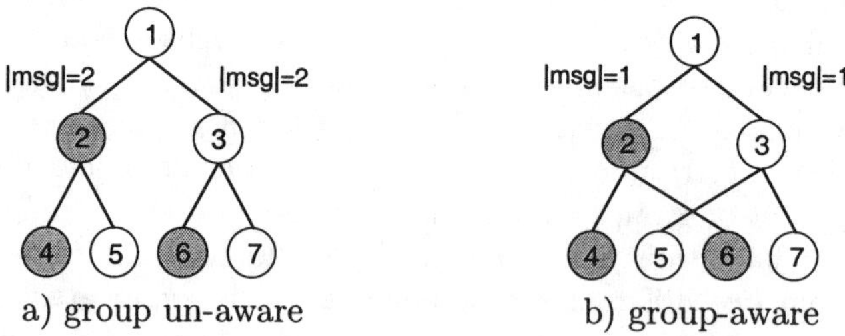

Figure 1: Benefits of group-aware network configuration.

4.2 GaNC Protocol

Our proposed *Group-Aware Network Configuration* method, or *GaNC*, constructs the routing tree as follows:

1. The root sensor prepares a query message which includes the query specification. The root sensor also sets the (L_s) value in the message to its level value (i.e., L_{root}, which is initially set to 0). It also sets the (G_s) to be its group id. It then broadcasts this query message to the neighboring sensors.
2. A sensor i that receives a query message and has its level value currently equal to ∞ will set its level to the level of the node it heard from, plus one. That is, $L_i = L_s + 1$.
3. Sensor i will also set its parent value P_i to Id_s and its parent's group id PG_i to G_s. It will then set Id_s, L_s and G_s in the query message to its own Id_i, L_i and G_i respectively and broadcast the query message to its neighbors.
4. While there are still query messages being propagated around the network, node i continues to listen to all messages it can hear.
5. If node i hears a message from a node at the same level as itself minus one $(L_i - 1)$, it uses *tie-breaker conditions* to decide if this new node should become its new parent. If so, node i makes Id_s its new parent.
6. Steps 2-5 are repeated until all query messages in the network have been sent out and received.

The GaNC algorithm is similar to the FHF algorithm. The main difference is that a child under the GaNC method can switch to a "better" parent while the tree is still being built. This switch is based on a set of *tie-breaker conditions* that go beyond the network characteristics and introduce the semantics of aggregation.

The goal of the the GaNC algorithm is to incorporate group identity into the routing tree construction. As such, the first tie-breaker condition (for Step 5 of the algorithm) is whether the child has the same group id as the parent. As long as a child is within listening distance of multiple parent choices, a child will choose a parent that has the same group id as itself instead of a parent from a different group. This is a choice that will allow parents and children to be in the same group as much as possible.

In the general case, a sensor node will be within listening range of multiple other nodes. Despite the savings in clustering nodes of the same group along the same path, a node that is far away will require significantly more transmission energy, and as such is not a good candidate. For that reason, we introduce a *distance factor*, *df*, that will limit the maximum range for which we consider candidate sensor nodes (for coming up with a "better" parent node). Under this approach, if d_i is the shortest distance seen so far (based on an estimation from signal strength), we will only consider nodes whose distance from a child node is at most $df \times d_i$, for example, for $df = 1.2$ we will only allow up to 20% more than the minimum distance.

Thus, the second tie-breaker is the estimated distance (or link quality) from the child to the parent. The parent with the lowest distance will be chosen in cases when there is more than one parent to choose from (that is in the same group as the child), or when no parents are in the same group as the child. The reason for this is that in both cases, routing through the closest parent will save transmission energy for the child.

4.3 GaNCi: GaNC Improvement

As an improvement on the original GaNC algorithm we also looked at allowing the child to choose a parent from a larger selection of nodes. In the original algorithm (Step 5 above) a child would consider a node as a possible parent if its level was that of the child minus one. The idea behind this constraint was that a child should always try to get one level closer to the root when choosing a parent. However, this limits the number of choices for selecting a parent and hence reduces the chances of the node finding a parent in the same group as itself.

In GaNCi (GaNC improved), the improvement we made was to change Step 5 to consider nodes that are in the same level as itself in addition to those in level lower than itself. In essence, the child can now choose a parent both from potential parents as designated in the original algorithm and from its own siblings. The benefits from this should be that more nodes have a better chance of having a parent in the same groups as themselves. This improvement should even surpass that of original GaNC, solely because there is a greater chance of children being in the same group as their parent.

In order to prevent siblings from selecting each other as their parent, which will prevent any information from those nodes or nodes routed through them from being propagated to the root, GaNCi requires each parent node to maintain a *child list*, much like the waiting list in Cougar. When a child chooses a new parent, it broadcasts a message letting all nearby sensors know of the change. Using the information from this message, both the previous parent (if one existed) and the newly identified parent for the child can update their child lists appropriately. This thereby accomplishes both tasks required by the new protocol: (1) allowing parents to create an up to date list of children it has and (2) allowing old parents to remove children that have switched to a new parent. To incorporate these changes into GaNCi, Step 5 of GaNC protocol is replaced with the following:

5 a. If node i hears a message from a node n at the same level as itself or lower ($L_n \leq L_i$) and n is a child of i and is announcing a new parent, remove n from child list of i.
b. If node i hears a message from a node n at the same level as itself or lower ($L_n \leq L_i$) and n is not a child of i, use *tie-breaker conditions* to decide if this new node should become the new parent and notify parent n that i has chosen it.

It is important to note that this improvement can be immediately incorporated into a scheme such as Cougar which is already keeping a waiting list of its children. In a scheme like TAG, the sensor node will have to be modified so that it can keep track of all known children.

5 EVALUATION TESTBED

In order to empirically study in-network aggregation in sensor networks, we created a simulation environment using CSIM [15]. Following typical sensor network simulation practices, the simulated network was configured as a grid of sensors. Each node could transmit data to sensors that were at most one hop away from it. In a grid this means it could only transmit to at most 8 other nodes. We simulated a contention-based MAC protocol (PAMAS) which avoids collision [17]. In this protocol, a sender node will perform a carrier sensing before initiating a transmission. If a node fails to get the medium, it goes to sleep and wakes up when the channel is free.

In the following experiments, we used the Group-By query format as described in Section 4, with the network producing results at intervals defined in the query. The size of the sensor networks is varied between 15x15 and 45x45, and the type of aggregation being performed is SUM. The group ids of the sensor nodes are based on randomness and the number of groups is varied between 2 and 50. These experiment parameters along with all other parameters are summarized in Table 1.

5.1 Random Walk

We used the random walk model for data generation. In this model, the domain of values was between 1 and 100 (to approximate temperature readings in Fahrenheit). A sensor reading is generated once at the beginning of each query interval. The value changes between one interval

Table 1: Simulation Parameters.

Parameter	Value	Default
Grid Size	15x15 – 45x45	15x15
Number of Groups	2 – 50	15
Aggregate	SUM	
TiNA *tct* amount	0% – 30%	
Number of Epochs	100	
Routing Scheme	FHF, GaNC	FHF
Randomness Degree	0.0 – 1.0	0.5
Random Step Size Limit	10%–50% of domain	10%

to the next with a probability known as the *randomness degree* (RD). Each time a sample is to be generated, a coin is tossed. If the coin value is less than RD, then a new value is generated, otherwise the sample value will be the same as before. For example, if $RD = 0.0$, then the value sampled by a sensor will never change, while if $RD = 0.5$, then there is a 50% chance that the new value at time t is different from the value at time $t+1$. We used the *Random Step Size Limit* to restrict how much the new value can deviate from the previous value. This limit is expressed as a percentage over the domain of values. In our case, a 10% limit implies that a new reading can differ by at most 10 (=10% of 100) compared to the previous reading.

5.2 Performance Metrics

In our experiments we focused on two measurements: energy consumption and relative error.

Energy: Energy is consumed in four main activities in sensor networks: transmitting, listening, processing, and sampling. We focused on transmission and listening power, since the amount of time spent sampling is the same for all techniques. We did not include energy required for processing because it is negligible compared to that needed for communication.

As mentioned before, a sensor node will send its data to the root through its assigned parent. A parent node is one hop away from its child, and one hop closer to the root than its child. So every node sends its data exactly one hop away, all of which are the same distance from one another. This allows us to assume a uniform cost of transmitting data. However, the overall energy consumed to transmit a partial result depends on the size of the partial results and the number of messages.

The values of the parameters needed to calculate the transmission cost were the same as in [4]. Specifically, we simulated sensors operating at 3 Volts and capable of transmitting data at a rate of 40 Kbps. The transmit current draw is 0.012 Amp while the receive current is 0.0018. Hence, the cost of transmitting one bit in terms of energy consumption units ($Joules$) is computed as:

T_{cost} = 3 Volt * 0.012 Amp * 1/40,000 Sec = 0.9 μJoules.

The cost of listening for one second is computed as:

R_{cost} = 3 Volt * 0.0018 Amp = 0.0054 Joules.

The energy consumed during listening is independent of the number of messages received by the sensor. It only depends on the time spent by the sensor being active and listening. Cougar does not specify when a sensor stops listening and switches to doze mode. Hence, in our simulation, we assumed that each sensor will start listening at the beginning of each round. After a sensor receives data from all nodes on its waiting list it will switch to doze mode.

Relative Error Metric: The *relative error metric* (REM) is a measure of how close the exact answer and the approximate answer are. The exact answer is generated if all sensors deliver their current readings within the each epoch interval. An approximate answer is one where some sensors decide not to send their reading. In our experiments, a sensor decides not to send a message because of the user-specified temporal coherency tolerance when TiNA is used.

In order to compute REM, we first need to measure the error over the Group-By query. We measured this error as described in [14]. Assume a query aggregates over a measure attribute M. Let $\{g_1, ..., g_n\}$ be the set of all groups in the exact answer to the query. Finally, let m_i and m_i' be the exact and approximate aggregate values over M in the group g_i. Then, the error ϵ_i in group g_i is defined to be the relative error, i.e., $\epsilon_i = \frac{(|m_i - m_i'|)}{m_i}$. The error δ over the Group-By query is defined as: $\delta = \frac{1}{n}\sum_{i=1}^{n} \epsilon_i$ finally, the average REM (or simply, REM) over time is defined as:

$$REM = \frac{1}{T}\sum_{t=1}^{T} \delta_t$$

where δ_t is the query error at $epoch_t$.

We are using REM as indication of the *quality of data* (QoD), where a high REM reflects a low QoD, while a low REM corresponds to a high QoD. Hence, we will be using both terms interchangeably.

6 EXPERIMENTS AND RESULTS

Using the simulator, we performed extensive experiments to evaluate the performance of our two algorithms GaNC and GaNCi. We compared GaNC and GaNCi with the FHF network configuration scheme presented in Section 3. Given that Cougar and TiNA require no modifications to use our algorithms, in this section we present our experiments with these two schemes.

6.1 Effects of Group-Aware Network Configuration

In this experiment we compare FHF with GaNC and GaNCi for varying *tct* amounts. We will show the effects of GaNC for network sizes of 15x15 and 45x45 and report both the energy savings and the quality of data.

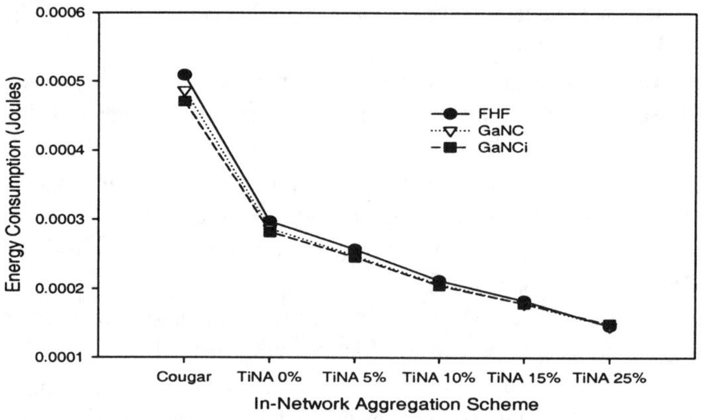

Figure 2: GaNC in 15x15 Grid (Energy).

Figures 2 and 3 show the energy and relative error for the 15x15 size network and Figures 4 and 5 show the same for the 45x45 size network. The number of groups used in this experiment was set at 15.

The first observation is that for the most part, using GaNC decreases the amount of energy used by the sensor network. This is especially prominent in larger networks. For TiNA with *tct=0%*, energy savings are 3.4% for the 15x15 grid and 23.6% for the 45x45 grid. The savings when using GaNC with plain Cougar are even higher: 4.4% for the 15x15 grid and 28.1% for the 45x45 grid. This shows that the proposed group-aware network configuration method can reduce energy significantly when used with Cougar and can even help save additional energy when used in conjunction with the TiNA framework.

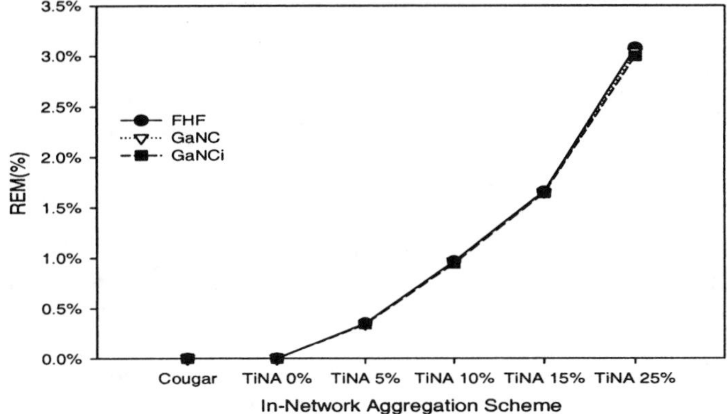

Figure 3: GaNC in 15x15 Grid (REM).

A related observation that can be made is about the relationship between GaNC and GaNCi for energy savings. By giving the child more choices for selecting a parent, there is an even better chance the child will fall into the same group as its parent. This should bring even greater energy savings, which the experiments show is true. For the *tct=0%* and grid size of 15x15, GaNCi saves 5.1%, an additional 1.7% over normal GaNC. For Cougar and the 15x15 grid, GaNCi saves 7.5%, an additional 3.1%. These additional savings are even greater for the 45x45 sized grid. For *tct=0%* and Cougar, GaNCi saves 26.7% and 32.3% respectively over FHF. These are additional savings of 3.1% and 4.2% over normal GaNC, showing that the GaNC improvement does help increase the amount of energy saved.

The next observation is that regardless of the size of the network, the energy savings of GaNC over FHF decrease as the *tct* of TiNA increases. In fact, for the 15x15 network, there is cross-over point where FHF requires the same amount of energy as GaNC and both require less energy than GaNCi. This is illustrated in Figure 2 where GaNC was saving 3.4% and GaNCi was saving 5.1% over FHF for *tct=0%*. When *tct* is increased to 25%, GaNC is using the same amount of energy as FHF and GaNCi is using 2% more energy than either FHF or GaNC. This result is not entirely surprising, however. When using GaNC some nodes will switch to parents that are in the same group as themselves. While this tends to decrease the number of message a parent sends, there are cases where switching parents can cause more messages to be sent. This is increased with GaNCi which has an even better chance of changing parents.

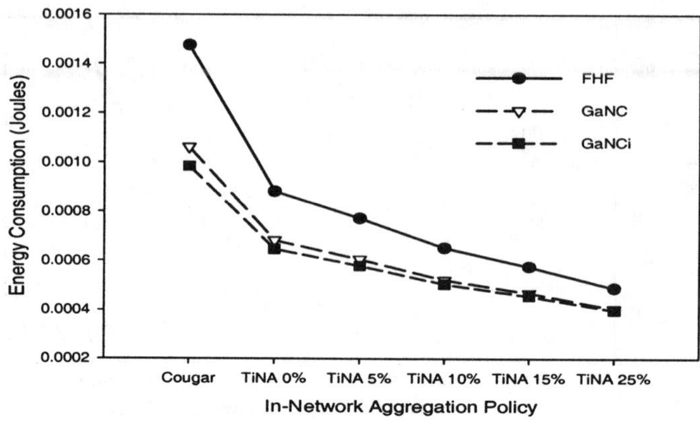

Figure 4: GaNC in 45x45 Grid (Energy).

For example, assume that, using FHF, two children of the same group are routed through a parent of a different group. This would result in 3 messages being sent overall (one from each child and one from the parent further up the tree). If, however, when GaNC is used, each of these children change to be with a parent of their same group, they may end up choosing two different parents, because the parents in their same group are not close enough to be "clustered" together. In order to propagate information up the tree under this setup, 4 messages are needed (one from each child and one from each parent. This is a 25% increase in the total number of messages, which in turn causes an increase on the energy used in the network.

For larger networks, the positive effects of using GaNC will outweigh the negative effects (per our previous example). As the *tct* increases, there are less nodes that are transmitting, since their value changes are not violating the specified *tct*. In larger networks, since there are many nodes, there will still be a lot of nodes transmitting, even under high *tcts*. In Figure 4, we can see that for the 45x45 grid, the energy savings at *tct=0%* are 23.6% and they drop to down to 17.1% for *tct=25%* (however, there is no cross-over point in this case).

The final observation is that there is very little difference in the relative error between using GaNC versus using FHF to create the routing tree, even with the large savings in energy (from GaNC). For the 15x15 grid, the relative error has decreased in the case of GaNC. This decrease is minor, ranging from .005% for *tct=5%* to .04% for *tct=25%*, but exists nonetheless. These decreases are even larger for GaNCi, where relative

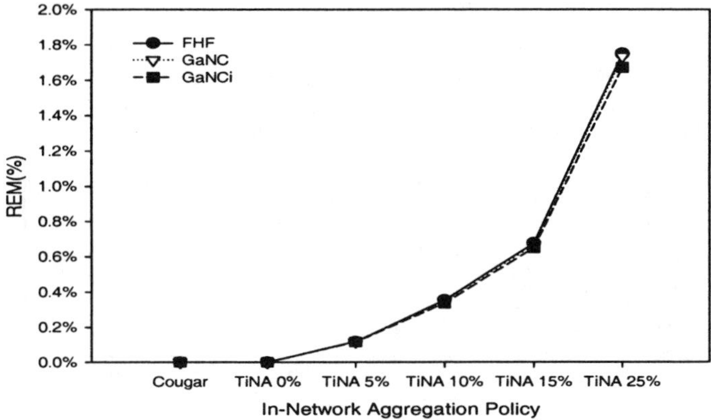

Figure 5: GaNC in 45x45 Grid (REM).

error has decreased between .01% for *tct=5%* to .08% for *tct=25%*. This improvement is due to the "free" ride some parent nodes may get by having a child already sending the same group as the parent and therefore be able to aggregate its own reading into the group aggregate without adding to the amount of energy used.

In the case of the 45x45 grid, the relative error again decreases by a small amount. This decrease was between .001% for *tct=5%* and .016% for *tct=25%*. Again, for GaNCi, the decrease is a little larger with decreases ranging from .002% for *tct=0%* to .077% for *tct=25%*. We have observed, however, that with smaller number of groups the relative error can also increase by very small amounts. This small increase is explained by multiple nodes counteracting each others' changes; thus the internal node does not get the "free" ride it may have gotten with only one child changing values.

6.2 Synergy of TiNA and GaNC under varying number of groups

In this experiment we examine how the number of groups affects the behavior of GaNC. Figure 6 shows the results from this experiment. We compared plain Cougar, TiNA over Cougar with *tct=0%*, and TiNA over Cougar with *tct=10%*. We ran two sets of experiments, one where FHF was used for configuring the network and one where we used GaNC instead (we note such cases with +GaNC). We do not show the runs with GaNCi in this experiment because the general pattern is the same as for GaNC, the only difference is the line would be slightly lower. We did this to make the figure easier to read.

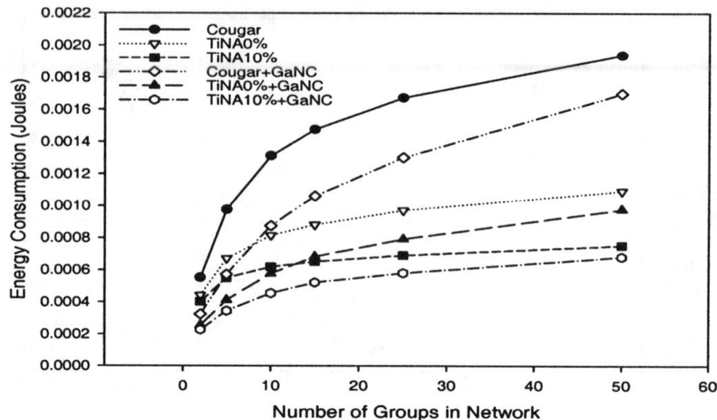

Figure 6: Energy Comparison for varying number of groups.

The number of groups in the experiment ranged from 2 to 50. We had a total of 45x45 = 2025 nodes. With just 5 groups, GaNC is expected to reduce energy consumption significantly, since children nodes will be able to select parents that are in the same group as them (in other words, GaNC will have a lot of options to choose from). For all three cases, using GaNC instead of FHF saves 41.4% in energy for Cougar, 38.5% for TiNA(0%), and 37.4% for TiNA(10%).

Using GaNC when the number of groups is 50 will not reduce energy as dramatically as with the case of 5 groups. When the number of groups is high, there are less chances that a child can find a parent in the same group. Based on the grid configuration (which we used in our simulations), each child has a maximum of three different parents to choose from. With 50 different groups, the chance that the child is in the same group as one of those three parents is less than 1%, so the savings will be minimal.

This can be observed from Figure 6: for 50 groups the savings with GaNC are only 12.4% for Cougar, 10.7% for TiNA(0%), and 9.5% for TiNA(10%). Overall, we see that when the number of groups increases, the savings with GaNC also decrease, but are still significant even when there are 50 different groups and the chance of a node actually switching parents in an efficient way is less than 1%.

7 RELATED WORK

The idea of exploiting the application semantics for data routing in sensor networks has been presented in [20, 9], where the goal is to use information-directed routing in order to minimize communication cost while maximizing information aggregation. The work in [9] showed the significant gains of applying information-directed routing in locating and tracking moving objects.

Using the query semantics for efficient data routing was introduced in [11], which proposed the use of a semantic routing tree (SRT). The basic idea that motivates the use of SRT is the fact that a given query does not apply to all nodes in the network. Hence, those nodes for which the query does not apply can be excluded from the query in order to save communication costs. As in GaNC, the work on SRT considered optimizing the routing tree for the special case of constant-valued attributes (e.g., location). However, the objective of the SRT is providing a design to minimize the number of nodes participating in a query with a predicate over that constant-valued attribute. Instead, in GaNC, the objective is to cluster along the same path sensor nodes that belong to the same group in order to maximize the benefit of in-network aggregation in reducing the size of messages, even if all nodes participate in the query.

8 CONCLUSIONS

In this work we explored the idea of *influencing the construction of the routing trees for sensor networks* with the goal of reducing the size of transmitted data in networks with in-network aggregation, hence providing additional energy efficiency in sensor networks. We proposed two network configuration algorithms for sensor networks, called *GaNC* and *GaNCi*, that achieve energy savings by considering the semantics of Group-By queries and the properties of the sensor nodes. These two algorithms work synergistically with existing in-network aggregation methods but differ with respect to their applicability. GaNC can be immediately combined with existing schemes such as Cougar and TAG. On the other hand, GaNCi which can be readily combined with schemes such as Cougar, may require additional features to work with schemes such as TAG.

We have shown experimentally that our algorithms result in large savings in energy over typical in-network aggregation methods. In addition, we looked at using GaNC and GaNCi in conjunction with TiNA which is another scheme for saving energy in-network aggregations methods. Our results have shown that GaNC can save up to 33% in energy over existing in-network aggregation schemes and can save an additional 29% over the savings of TiNA, when used in tandem with it. We also showed that using GaNC and GaNCi with TiNA can help improve the quality of data in systems with imperfect quality of data.

REFERENCES

[1] P. Bonnet, J. Gehrke, and P. Seshadri. Towards sensor database systems. In *Proc. of MDM*, 2001.

[2] J. Heidemann, F. Silva, C. Intanagonwiwat, R. Govindan, D. Estrin, and D. Ganesan. Building efficient wireless sensor networks with low-level naming. In *Proc. of SOSP*, October 2001.

[3] W. Heinzelman, A. Chandrakasan, and H. Balakrishnan. Energy-efficient communication protocol for wireless microsensor networks. In *HICSS*, January 2000.

[4] J. Hill and D. Culler. Mica: A wireless platform for deeply embedded networks. *IEEE Micro.*, 22(6), 2002.

[5] J. Hill, R. Szewczyk, A. Woo, S. Hollar, D. Culler, and K. Pister. System architecture directions for networked sensors. In *Proc of ASPLOS*, 2000.

[6] C. Intanagonwiwat et al. Directed diffusion: A scalable and robust communication paradigm for sensor networks. In *Proc. of MOBI-COM*, August 2000.

[7] P. Juang et al. Energy-efficient computing for wildlife tracking: design tradeoffs and early experiences with zebranet. In *Proc. of ASPLOS'02*.

[8] C. Lin, C. Federspiel, and D. Auslander. Multi-sensor single actuator control of hvac, 2002.

[9] J. Liu, F. Zhao, and D. Petrovic. Information-directed routing in ad hoc sensor networks. In *2nd ACM international conference on Wireless sensor networks and applications*, 2003.

[10] S. Madden, M. Franklin, J. Hellerstein, and W. Hong. TAG: a tiny aggregation service for ad-hoc sensor networks. In *Proc. of OSDI*, 2002.

[11] S. Madden, M. Franklin, J. Hellerstein, and W. Hong. The design of an acquisitional query processor for sensor networks. In *Proc. of ACM SIGMOD*, 2003.

[12] A. Mainwaring, J. Polastre, R. Szewczyk, D. Culler, and J. Anderson. Wireless sensor networks for habitat monitoring. In *Proc. of ACM WSNA'02*, 2002.

[13] C. Perkins. Ad-hoc on demand distance vector routing (AODV). Internet-draft, November 1997.

[14] S.Acharya, P. B. Gibbons, and V. Poosala. Congressional samples for approximate answering of group-by queries. In *Proc of ACM SIGMOD*, 2000.

[15] H. Schwetman. CSIM user's guide. MCC Corp.

[16] M. A. Sharaf, J. Beaver, A. Labrinidis, and P. K. Chrysanthis. Tina: A scheme for temporal coherency-aware in-network aggregation. In *Proc. of MobiDE*, 2003.

[17] S. Singh and C. Raghavendra. PAMAS: Power aware multi-access protocol with signalling for ad hoc networks. *ACM Computer Comm. Review*, 28(3).

[18] A. Woo and D. Culler. A transmission control scheme for media access in sensor networks. In *ACM Mobicom*, July 2001.

[19] Y. Yao and J. Gehrke. Query processing for sensor net. In *Proc. of CIDR*, 2003.

[20] F. Zhao, J. Shin, and J. Reich. Information-driven dynamic sensor collaboration. *IEEE Signal Processing Magazine*, vol. 19, 2002.

Synthetic Data Generation to Support Irregular Sampling in Sensor Networks

Yan Yu[†] Deepak Ganesan[†] Lewis Girod[†]
Deborah Estrin[†] Ramesh Govindan[†‡]

†Department of Computer Science, University of California at Los Angeles
Los Angeles, CA 90095

†‡ Computer Science Department, University of Southern California / ISI
Los Angeles, CA 90089

ABSTRACT

Despite increasing interest, sensor network research is still in its initial phase. Few real systems have been deployed and little data is available to test proposed protocol and data management designs. Most sensor network research to date uses randomly generated data input to simulate their systems. Some researchers have proposed using environmental monitoring data obtained from remote sensing or in-situ instrumentation. In many cases, neither of these approaches is relevant, because they are either collected from regular grid topology, or too coarse grained. This paper proposes to use synthetic data generation techniques to generate irregular data topology from the available experimental data. Our goal is to more realistically evaluate sensor network system designs before large scale field deployment.

Our evaluation results on the radar data set of weather observations shows that the spatial correlation of the original and synthetic data are similar. Moreover, visual comparison shows that the synthetic data retains interesting properties (*e.g.*, edges) of the original data. Our case study on the DIMENSIONS system demonstrates how synthetic data helps to evaluate the system over an irregular topology, and points out the need to improve the algorithm.

1 INTRODUCTION

Despite increasing interest, sensor network research is still in its initial phase. Few real systems are deployed and little data is available to test proposed protocol designs. Most sensor network research to date uses randomly generated data input to evaluate systems. Evaluating the system with data representing real-world scenarios or representing a wide range of conditions is essential for systematic protocol design and evaluation of sensor network systems whose performance is sensitive to the spatio-temporal features of the system inputs. To our knowledge, there has been no previous work done on modeling data input in a sensor network context.

Some researchers proposed using environmental monitoring data obtained from remote sensing or in-situ instrumentation. However, these data are mostly collected from a regular grid configuration. Due to the large scale deployment, the proposed sensor networks (*e.g.*, in habitat monitoring [6]) are most likely in an irregular topology. Further, the granularity and density of those data sets does not match the expected granularity and density of future sensor network deployment. Although they cannot be directly used to evaluate the sensor network algorithms, they can provide useful models of spatial and temporal correlations in the experimental data, which can be used to generate synthetic data sets. Because many sensor network protocols exploit spatial correlations, we are interested in synthetic data that have similar spatial correlations as that of the experimental data. In this paper we focus on modeling the experimental data to generate irregular topology data for two reasons: First, we lack ground truth data to verify that the synthetic data match some interesting statistics of the experimental data at the scale of fine granularity. Second, we cannot assume that the experimental data are generated from a band-limited spatial process.

In order to evaluate sensor network algorithms under different topologies other than the single topology associated with the available data set, we proposed to generate irregular topology data. We first apply spatial interpolation techniques, implicitly or explicitly model the spatial and temporal correlation in a data set. From this empirical model, we generate ultra fine-grained data, and then use it to generate irregular data. This technique will also allow us to study system performance under various topology, but with the same data correlation model. On the other hand, by using the same experimental data setting, and plugging in different correlation models, we are able to evaluate how the algorithms interact with various data correlation characteristics. In this paper, we use the DIMENSIONS [12] system as our case study and investigate the impact of irregular topologies on algorithm performance. DIMENSIONS provides a unified view of data handling in sensor networks, incorporating long-term storage, multi-resolution data access and spatio-temporal pattern mining. It is designed to support observation, analysis and querying of distributed sensor data at multiple resolutions, while exploiting spatio-temporal correlation. While the interplay of topology and radio connectivity has been studied in-depth in the context of sensor networks (*e.g.*, ASCENT [7], GAF/CEC [30], STEM [25] etc.) there is little work on studying the interplay between in-network data processing and topology. Our models and synthetic data sets are intended to help study the coupling between the topology and data processing schemes in such networks.

In the remainder of this paper, we first review related work in section 2. In section 3, we start with how to generate fine grained spatial data maps using a model of spatial correlation as well as how to generate fine grained spatio-temporal data sets using a joint space-time model. This is an essential step in irregular data generation, which we discuss in section 4. We also present results of applying these two modeling techniques to an experimental radar data set in section 3. In section 4, we use the DIMENSIONS system as a case study to demonstrate how the synthetic data from the modeling of experimental data helps in system evaluation, and point out the need to improve the algorithm. We conclude in section 5.

2 RELATED WORK

Data modeling techniques in environmental science To the best of our knowledge, no previous work has been done on data modeling in a sensor network context. However, in environmental science or geophysics, various data analysis techniques have been applied to extract interesting statistical features from the data, or estimate the data values at un-sampled or missing data points. Various spatial interpolation techniques, such as Voronoi polygons, triangulation, naturall neighbor interpolation, trend surface or splines [28], have been proposed. Kriging, which refers to a family of generalized least-squares regression algorithms, has been used extensively in various environmental science disciplines. Kriging models the spatial correlation in the data and minimizes the estimation variance under the unbiasedness constraints of the estimator. In this paper, we reported our experience with Kriging and several non-stochastical interpolation techniques.

In addition, there is significant research devoted to time series analysis. Autoregressive integrated moving average model (ARIMA) [3] explicitly considers the trend and periodic behavior in the temporal data. The wavelet model [11] has been successfully used to model the cyclic, or repeatable behavior in data. In addition, researchers have also explored neural networks [9], kernel smoothing for time series analysis.

Joint spatio-temporal models have received much attention in recent years [17, 24, 23, 19] because they inherently model the correlation between the temporal and spatial domain. The joint space-time model used in our data analysis is inspired by and simplified from the joint space-time model proposed by Kyriakidis *et al.* [18]. In [18], co-located terrain elevation values are used to enhance the spatial prediction of the coefficients in the temporal model constructed at each gauge station. However, this requires the availability of an extra environmental variable, which does not exist in our case.

Data modeling in Database and Data Mining Theodoridis *et al.* [26] proposes to generate spatio-temporal datasets according to parametric models and user-defined parameters. However, the design space is huge, it is impossible to exhaustively visit the entire design space, *i.e.*, generate data sets for every possible set of parameter values. Without additional knowledge, we have no reason to believe that any parameter setting is more realistic or more important than others. Therefore we proposed to start with an experimental data set, and generate synthetic data that shares similar statistics with the experimental data.

Given a large data set that is beyond the computer memory constraints, data squashing [27] proposes schemes to shrink a large data set to manageable size. Although sharing the same objective of deriving synthetic data from modeling existing data as we do, they consider non-spatio-temporal datasets. The spatio-temporal data cannot be assumed to be drawn from the same certain probability model as assumed by [27].

TCP traffic Modeling in Internet In a similar attempt to model the data input to the network system in an Internet context, researchers have studied TCP traffic modeling. For example, Caceres *et al.* [5] characterized and built empirical models of wide area network applications. The specific data modeling technique in their study [5] does not apply to sensor networks due to the following: (a) Sensor networks are closely coupled with the physical world, therefore the data modeling in sensor networks needs to capture the spatial and temporal correlation in a highly dynamic physical environment. (b) The characteristics of wide area TCP traffic is potentially very different from the workload or traffic in sensor networks.

System components modeling in wireless ad-hoc networks and sensor networks Previous research has been carried out on modeling system components in ad-hoc networks and sensor networks, however, to our knowledge, none of this research has focused on modeling the data input to the system.

Among the work on modeling system components in the context of ad-hoc networks, [4, 8] use regular or uniform topology setups, and "random waypoint" models in their protocol evaluations, and [22, 14] discuss multiple topology setups and mobility patterns for more realistic scenarios. In modeling wireless channels, Konrad *et al.* [15] study non-stationary behavior of packet loss in the wireless channel and modeled the GSM (Global System for Mobile) traces with a Markov-based Trace Analysis (MTA) algorithm.

Ns-2[2] and GloMoSim [32] provide flexibility in simulating various layers of wired networks or wireless ad-hoc networks. However, they do not capture many important aspects of sensor networks, such as sensor models, or channel models. In contrast, Sensorsim [20, 21] directly targets sensor networks. In addition to a few topology and traffic scenarios, they introduce the notion of a sensor stack and sensing channel. The sensor stack is used to model the signal source, and the sensing channel is used to model the medium which the signal travels through. Our work could be used as a new model in Sensorsim.

3 SYNTHETIC DATA GENERATION BASED ON EMPIRICAL MODELS OF EXPERIMENTAL DATA

Before delving into irregular topology data generation, we start with the problem of generating fine-grained synthetic data, which is an essential step in our irregular topology data generation. Our proposed synthetic data generation includes both spatial and spatio-temporal data types. To generate spatial data, we start with an experimental data set which is a collection of data measurements from a study area. Assuming the data is a realization of an ergodic and local stationary random process, we use spatial interpolation techniques to generate synthetic data at unmonitored locations.

Similarly, to generate synthetic spatio-temporal data, we again start with an experimental space-time data set, which includes multiple snapshots of data measurements from a study area at various times. If we were only interested in data at recording time, we could apply our proposed spatial interpolation techniques to each snapshot of data separately, then generate a collection of spatial data sets at each recording time. However, this does not allow us to generate synthetic data at times other than the recording times. In addition, the joint space-time correlation is not fully modeled and exploited if we model each snapshot of spatial data separately. Therefore, we propose to model the joint space-time dependency and variation in the data. Inspired by a joint space-time model in [18], we model the data as a joint realization of a collection of space indexed time series, one for each spatial location. Time series model coefficients are space-dependent, and so we further spatially model them to capture the space-time interactions. Synthetic data are then generated at unmonitored locations and time from the joint space-time model. This allows us to generate synthetic data at arbitrary spatial and temporal configurations.

In the remainder of this section, we first discuss spatial interpolation techniques and present the results of radar dataset applications . Then we discuss a joint spatio-temporal model and the result of applying it to the same radar data set.

3.1 Generating Synthetic Spatial Data Sets We start with an experimental data set, which is typically sparsely sampled. To generate a large set of samples at much finer granularity, a spatial interpolation algorithm is used to predict at unsampled locations. The spatial interpolation problem has been extensively studied. Both stochastic and non-stochastic spatial interpolation techniques exist, depending on whether we assume the observations are generated from a stochastic random process. In general, the spatial interpolation problem can be formulated as: Given a set of observations $\{z(k_1), z(k_2), ..., z(k_n)\}$ at known locations k_i, $i = 1, ..., n$, spatial interpolation is used to generate prediction at an unknown location u. However, if we take a stochastic approach, the above spatial interpolation problem can be formulated as the following estimation problem. A random process, Z, is defined as a set of dependent (here spatially dependent) random variables $Z(u)$, one for each location u in the study area A, denoted as $\{Z(u), \forall u \in A\}$. Assuming Z is an ergodic process, the problem is defined to estimate some statistics (*e.g.*, mean) of $Z(u)$ ($u \in A$) given a realization of $\{Z(u)\}$ at locations u_i, $i = 1, ..., n$, $u_i \in A$. A lies in one dimensional or high dimensional space.

Kriging [13] is a widely used geostatistics technique to address the above estimation problem. Kriging, which is named after D. G. Krige [16], refers to a range of least-squares based estimation techniques. It has both linear and non-linear forms. In this paper, ordinary kriging, which is a linear estimator, is used in our spatial interpolation and joint spatio-temporal modeling example.

In ordinary kriging, at an unmonitored location, the data is estimated as a weighted average of the neighboring samples. There are different ways to determine the weights, *e.g.*, assign all of the weight to the nearest data, as used in the nearest neighbor interpolation approach; assign the weights inversely proportional to the distance from the location being estimated. Assuming the underlying random process is locally stationary, Kriging uses a variogram[1] to model the spatial correlation in the data. The weights are determined by minimizing the estimation variance, which is written as a function of the variogram (or covariance). In addition to providing least squares based estimate, Kriging also provides estimation variance, which is one of the important reasons that Kriging has been popular in geostatistics. However, as we will explain shortly, estimation is not our ultimate goal; our goal is to generate fine grained sensing data which can be used to effectively evaluate sensor network protocols. Therefore we also study other non-stochastic spatial interpolation algo-

[1]Please refer to Appendix A for a brief introduction to variograms.

rithms: Nearest neighbor interpolation, Delaunay triangulation interpolation, Inverse-distance-squared weighted average interpolation, BiLinear interpolation, BiCubic interpolation, Spline interpolation, and Edge directed interpolation [10]. Due to space limit, please refer to [31] for details on the above spatial interpolation algorithms.

3.1.1 Evaluation of synthetic data generation

Data set description To apply the spatial interpolation techniques described above, we consider the resampled S-Pol radar data provided by NCAR[2], which records the intensity of reflectivity in dBZ, where Z is proportional to the returned power for a particular radar and a particular range. The original data were recorded in the polar coordinate system. Samples were taken at every 0.7 degrees in azimuth and 1008 sample locations (approximately 150 meters between neighboring samples) in range, resulting in a total of 500 x 1008 samples for each 360 degree azimuthal sweep. They were converted to the Cartesian grid using the nearest neighbor resampling method. A grid point is only assigned a value from a neighbor when the neighbor is within 1km and 10 degree range. If none of its neighbors are within this range, the grid point is labeled as missing value, *e.g.*, the NaN value is assigned. Resampling, instead of averaging, was used to retain the critical unambiguous and definitive differences in the data. In this paper, we select a subset of the data that has no missing values to perform our data analysis. Specifically, each snapshot of data in our study is a 60 x 60 spatial grid data with 1km spacing.

Spatial interpolation algorithms implementation We apply the above eight interpolation algorithms to the selected spatial radar data sets. We use the *spatial* package in R [1] to achieve Kriging. Nearest neighbor, Bilinear, Bicubic, Spline interpolation results were obtained from the interp2() function in Matlab. Since Bilinear and Bicubic interpolations provide no prediction for edge points, we use results from Nearest Neighbor interpolation for edge points in bilinear or bicubic interpolation. Edge directed interpolation is based on [10]. Inverse-distance-squared weighted average interpolation, and Delaunay triangulation interpolation were implemented in Matlab following the interface of interp2(). The *spatial* package in R and the interp2() function in Matlab generate output for a grid region. This motivates us to use the resampled grid data,

[2] S-Pol (S band polar metric radar) data were collected during the International H2O Project (IHOP; Principal Investigators: D. Parsons, T. Weckwerth, et al.). S-Pol is fielded by the Atmospheric Technology Division of the National Center for Atmospheric Research. We acknowledge NCAR and its sponsor, the National Science Foundation, for provision of the S-Pol data set.

instead of the raw data from the polar coordinate system.

Evaluation metrics For our synthetic data generation, we are interested in how close the synthetic data can approximate the interesting statistical features of the original data. The set of statistical features selected as evaluation metrics should be of interest to the algorithm and applications for which the synthetic data are intended to be used. It is hard to define a statistical feature set that is generally applicable to most algorithms and data sets, nevertheless, quite a few existing sensor network protocols (including DIMENSIONS, which is used as our case study) exploit spatial correlations in the data. In general, since sensor networks are envisioned to be deployed in the physical environment and deal with data from the geometric world, we believe that many sensor network protocols will exploit spatial correlation in the data. Therefore, besides visual comparison, we use spatial correlation (which is measured by its variogram values) of the synthetic data versus original data to assess the applicability of this synthetic data generation technique to the sensor network algorithm being evaluated. Suppose two data sets A and B, and their variogram values are $\{\hat{\gamma_1}(h_i)\}$ and $\{\hat{\gamma_2}(h_i)\}$ respectively, where h_i are sample separation distances between two observations; i =1, ..., m. The Mean Square Difference of variogram values of two data sets is defined as: $\sum_{i=1}^{m}(\hat{\gamma_1}(h_i) - \hat{\gamma_2}(h_i))^2$.

Interpolation resolution We studied two extremes of interpolation resolutions: (1) Coarse grained interpolation, in which case, we start from the down-sampled data (which reduces the data size in half in each dimension), increase the interpolation resolution by 4, compare the variogram value of the interpolated data with that of the original data. Note that the original data can be considered as ground truth in this case. The coarse grained interpolation is used to evaluate how the synthetic data generated by different interpolation algorithms approximate the spatial correlation of the experimental data. (2) Fine grained interpolation. Starting with a radar data set with 1km spacing, we increase the resolution by 10 times in each dimension, resulting in a 590x590 grid with 100m spacing. Fine grained interpolation is an essential step in generating irregular topology data.

Evaluation Results First we visually present how the spatial correlation (*i.e.*, variogram values) of the synthetic data approximates that of the original data in the case of coarse-grained interpolation. For the spatial dataset shown in Figure 1, Figure 2 shows the variogram plot of several synthetic data sets (generated from various interpolation algorithms) *vs.* that of the original data. It demonstrates that the variogram curves of most synthetic data (except the one

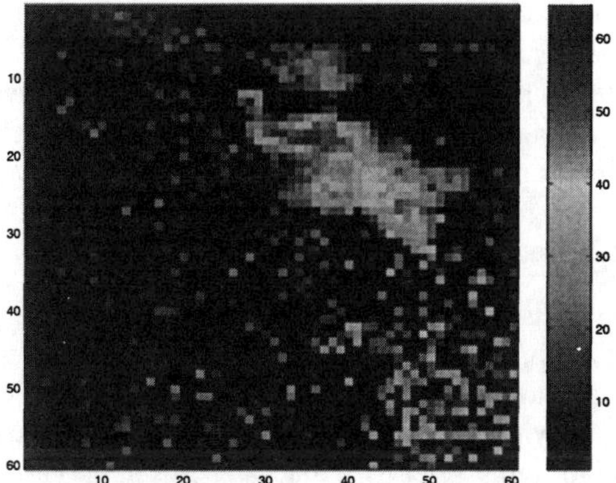

Figure 1: Spatial modeling example: original data map (60x60)

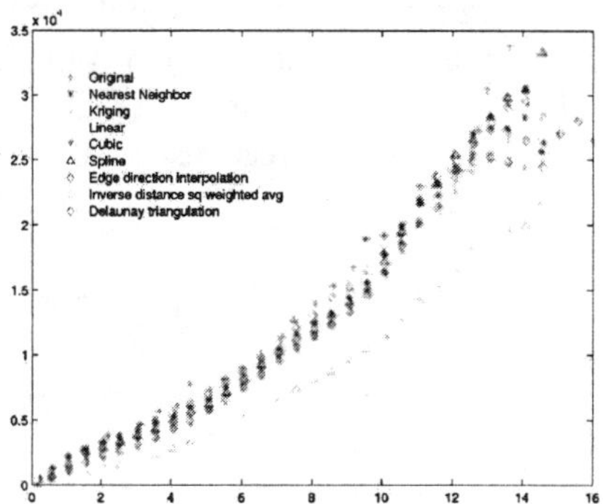

Figure 2: MSD of variogram values: Coarse grained interpolation results on a snapshot of radar data

from inverse-distance-squared weighted average interpolation) closely approximate that of the original one. At the long lag distances, the synthetic data may appear to slightly undereestimatethe long-range dependency in the original data. The source of this under-estimate may be due to the smoothing effect of the interpolation algorithms.

Further, we use the mean square difference between the variogram values of the original data and the synthetic data as a quantitative measure of how closely the synthetic data approximates the original data in terms of variogram values.

Table 1 lists the mean square difference results averaged over 100 snapshots of radar data in increasing order. For this radar data set, the nearest neighbor interpolation best matches with the original variogram, the inverse-distance-squared weighted averaging appeared the worst in preserving the original variogram, while the order of other interpolation algorithms changes between two different interpolation resolutions. We observe the same inconsistency with another precipitation data set [29].

Based on these results we do not recommend one single interpolation algorithm over others, but propose using spatial correlation as the evaluation metric for our synthetic data generation purpose and a suite of interpolation algorithms. Given a new synthetic data generation task, we would test with different interpolation algorithms, select one that can best suit the current application and experimental data set at hand. Note that although the Nearest neighbor interpolation appears best matching with the original variogram model, it is not appropriate in the case of ultra-fine grained interpolation, since it assigns all nodes in a local neighborhood the same value from the nearby sample. However, most physical phenomena have some degree of variation even in a small local neighborhood, and thus we would not expect all sensors deployed in a local neighborhood report the same sensor readings as in the case of the nearest neighbor interpolation.

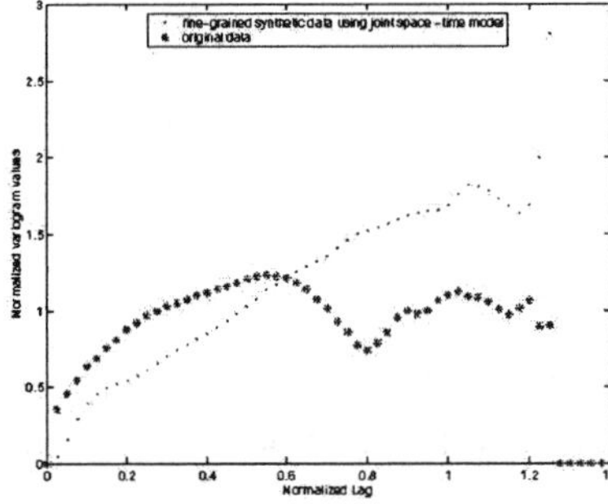

Figure 3: Spatial modeling example: Variogram of the fine-grained synthetic data and the original data.

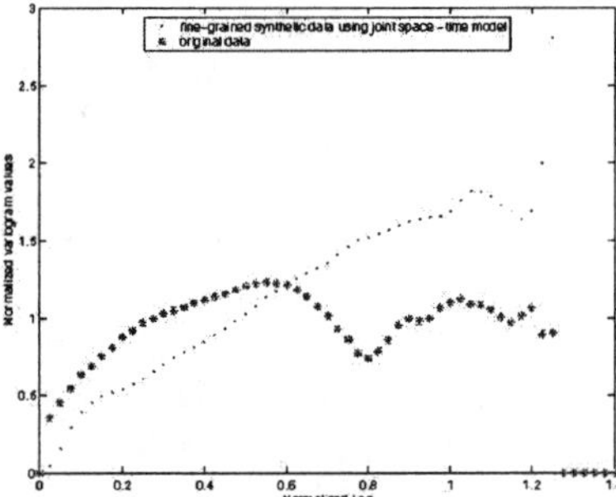

Figure 4: Joint modeling example: Variogram of the fine-grained synthetic data and the original data.

Summary: As shown above, most interpolation algorithms can approximate the original variogram models. However, it can only be used to interpolate at unsampled locations, not unsampled time. Furthermore, spatial interpolation algorithms, including Kriging, is not able to characterize the correlation be-

Name of method	MSD for coarse-grained interpolation
Nearest neighbor	8.354218e+01 (1.836358e+01)
Edge directed	1.970850e+02 (2.129320e+01)
Cubic	2.000790e+02 (1.694163e+01)
Delaunay triangulation	3.406270e+02 (4.795614e+01)
Linear	3.941510e+02 (2.876476e+01)
Spline	7.148526e+02 (5.5949e+01)
Kriging	1.469954e+03 (1.913371e+04)
Inverse-dist.-squared-weighted avg.	1.682726e+03 (3.617214e+02)

Table 1: Mean Square Difference of variogram values for different interpolation algorithms in the increasing order of MSD for coarse-grained interpolation. Here we use median from 100 snapshots instead of mean to get rid of outliers, and list 95% confidence interval in the brackets.

tween the spatial domain and temporal domain of the data, such as how the time trend varies at each location and how the spatial correlation changes as time progresses. Next, we wish to use the joint space-time model to address the limitations of the spatial interpolation techniques alone.

3.2 Joint space-time model When considering the time and space domains together, a spatial-temporal random process was often decomposed into a mean component modeling the trend, and a random residue component modeling the fluctuations around the trend in both the time and space domains. Formally,

$$Z(u_\alpha, t_i) = M(u_\alpha, t_i) + R(u_\alpha, t_i) \tag{1}$$

where $Z(u_\alpha, t_i)$ is the attribute value under study, u_α is the location, t_i is the time, $M(u_\alpha, t_i)$ is the trend, and $R(u_\alpha, t_i)$ is the stationary residual component.

For the trend component, we borrowed the model from [18] where Kyriakidis *et al.* built a space-time model for daily precipitation data in northern California coastal region. $M(u_\alpha, t_i)$, in Equation 1 is further modeled as the sum of $(K + 1)$ basis functions of time, $f_k(t_i)$: $M(u_\alpha, t_i) = \sum_{k=0}^{K} b_k(u_\alpha) f_k(t_i))$ where $f_k(t_i)$ is a function solely dependent on time t_i, with $f_0(t_i) = 1$ by convention. $b_k(u_\alpha)$ is the coefficient associated with the k-th function, $f_k(t_i)$, which is solely dependent on location u_α. $B(u_\alpha)$ and $F(t_i)$ can be computed as follows.

We first describe the guidelines to compute $f_k(t_i)$. [18] suggested that any temporal periodicities in the data should be incorporated in $f_k(t_i)$. Alternatively, $f_k(t_i)$ could also be identified as a set of orthogonal factors from empirical orthogonal function (EOF) analysis of the data, or the spatial average of data at a time snapshot. In this paper, we use two basis functions: $f_0(t_i) = 1$ by convention; for $f_1(t_i)$, we take the spatial average of each time snapshot of the data. Formally, $F(t_i)$ (for illustration convenience, we write $f_k(t_i)$ and $b_k(u_\alpha)$ in matrix formats) can be written as:

$$\begin{pmatrix} 1 & \frac{1}{n}\sum_u z(u,t_1) \\ 1 & \frac{1}{n}\sum_u z(u,t_2) \\ .. & ... \\ 1 & \frac{1}{n}\sum_u z(u,t_i) \end{pmatrix} \tag{2}$$

Next let us see $B(u_\alpha)$. If we ignore the residue component in Equation 1 for now, $z(u_\alpha, t_i)$ can be written as

$$Z(u_\alpha, t_i) = B(u_\alpha) \cdot F(t_i) \tag{3}$$

The vector of coefficients $B(u_\alpha)$ can be written as a weighted linear combination of the data vector $Z(u_\alpha, t_i)$: $B(u_\alpha) = H(u_\alpha) \cdot Z(u_\alpha)$, where $H(u_\alpha)$ is a matrix of weights assigned to each data component of $Z(u_\alpha)$ and $Z(u_\alpha)$ is a vector consisting of a time series data at location u_α. If the matrix F is of full rank, we have $H = (F' \cdot F)^{-1} \cdot F'$ from the ordinary least squares analysis (OLS).

The joint spatio-temporal trend model is constructed at each monitored location. The resulting trend parameters, $\{b_k(u_\alpha)\}$, are spatially correlated since they are derived from the same realization of the underlying spatio-temporal random process. Therefore, we spatially model and interpolate the trend parameters, $\{b_k(u_\alpha)\}$, using Kriging (Note that other spatial interpolation techniques could also be used) to obtain the value of $\{b_k(u_\alpha)\}$ at unsampled location u_α. Similarly, $\{F_{t_i}\}$ can be modeled and interpolated to obtain the value of F_t at unsampled time point t.

3.2.1 Evaluation of joint space-time modeling

To apply the joint space-time model described above, we considered a subset of the S-Pol radar data provided by NCAR. We selected a 70 x 70 spatial subset of the original data with 1km spacing, and 259 time snapshots across 2 days in May 2002. As mentioned above, the synthetic data is desired to have similar spatial correlation as the original data. Here we use one snapshot to shed some light on how the synthetic data generated from the joint space-time model captures the spatial correlation in the original data. Figure 4 shows the variogram plot of the synthetic data (which is generated from the joint space-time model) *vs.* original data, where the variogram value is normalized by the variance of the data, and the lag distance[3] is normalized by the maximum distance in horizontal or vertical directions. For comparison, we also show the results from spatial Kriging for the same snapshot in Figure 3. In Figure 4, the variogram of the synthetic data approximates that of the original one in the range [0, 0.6], which is less than half of the maximum distance between any two nodes; while in the spatial interpolation case (Figure 3), similar trends between two variogram curves (of synthetic vs. original data) are observed except in lags with range [1.1, 1.3], where the last few points in the variogram may be less accurate due to the fact that it is based on less data compared to other portions of the variogram. The larger discrepancy between the synthetic and original data in Figure 4 suggests that the joint space-time model does not capture the spatial correlation in the original data as precisely as the purely spatial interpolation does.

Spatial modeling vs. Joint space-time model A joint spatio-temporal model can capture the correlation between the temporal trend and spatial variation and generate synthetic data at unmonitored times and locations. Further, compared with spatial interpolation at each time snapshot, joint space-time modeling is faster when we generate large amounts of snapshots of spatial data. For example, the precipitation data set [29] to be introduced next in section 4, includes daily precipitation data from the Pacific Northwest for approximately *45* years, or *16801* days. If we interpolate each daily snapshot separately, we would have to apply spatial interpolation technique 16801 times. However, using the joint space-time model described above, we need only to interpolate the coefficients $\{B_k(u_\alpha)\}$ $(k = 1, 2)$ in the spatial domain. This reduces the application of spatial interpolation to only two times and interpolating $F(t)$ in the temporal domain to only once.

The joint space-time model come at a cost. It combines spatial and temporal domains that have completely different characteristics. Therefore, a joint space-time model is usually more complicated and typically results in greater error than a space model. When we compare the prediction accuracy of spatial interpolation to the joint space-time model at a fixed point in time, a spatial interpolation technique usually can more closely capture the original data than a joint space-time model. In addition, spatial interpolation is often used as a component in the joint spatio-temporal models. Therefore, even though a joint space-time model naturally better captures the temporal and spatial characteristics in the data, the spatial modeling and interpolation is still important and popular in practice.

[3]Lag distance is defined as the distance between two points

4 CASE STUDY: USING SYNTHETIC DATA TO BETTER EVALUATE A SENSOR NETWORK PROTOCOL

In this section, we use DIMENSIONS [12] as an example of how synthetic data generated from empirical models aids in evaluating the performance of an algorithm. In particular, when experimental data sets are available only from regular grid topology, we show how synthetic data generation allows us to evaluate DIMENSIONS over irregular topologies. Even though we use it as a case study, our proposed approach to irregular topology data generation is by no means tied to the DIMENSIONS system. The irregular topology data generated from the procedure described next can be used to evaluate other sensor network algorithms.

1. Generating ultra fine-grained data: In Step 1 we create a grid topology at a much finer granularity than our target topology. We model the correlation in the experimental data using the joint spatio-temporal model described in Section 3.2, and further generate a much finer grained data set based on this empirical model. In this case study, we consider a rainfall data set that provides 50km resolution daily precipitation data for the Pacific NorthWest from 1949-1994 [29]. The spatial setup comprises a 15x12 grid of nodes, each node recording daily precipitation values. Since the data set covers 45 years of daily precipitation data, it is rich enough in the temporal dimension. Thus, we increase the data granularity only in the spatial dimension, leaving the granularity of the temporal dimension unchanged. This fine-grained model was used to interpolate 9 points between every pair of points in both horizontal and vertical dimension, thus resulting in a 140 x 110 grid data.

2. Creating irregular topology data: In Step 2 our objective is to downsample the fine-grained data set from Step 1, and generate a data set for an arbitrary topology. We overlay the target topology on this ultra fine-grained grid data. Each node in the target topology is assigned a value from the nearest grid data. Note that our joint space-time model could be directly used to generate synthetic data at an arbitrary location and time. However, providing an ultra fine-grained data set allows the protocol designers to derive arbitrary topology data as they wish from the fine-grained data, simplifying the generation of synthetic data in their chosen configuration.

   To create a random topology with a predefined number of nodes, we select grid points at random from the fine-grained grid data. In our case study, 2% of the nodes in the 140 x 110 fine-grained grid were chosen at random (i.e., each node was chosen with probability 2%). This results in approximately 14x11x2 nodes being chosen in the network.

Next we demonstrate how the synthetic data generated from the above procedure help to expose problems, and gain more insights into the current DIMENSIONS system design.

DIMENSIONS We now return to our case study to illustrate how data processing algorithms can be sensitive to topology features. DIMENSIONS [12] proposes wavelet-based multi-resolution summarization and drill-down querying. Summaries are generated in a multi-resolution manner, corresponding to different spatial and temporal scales. Queries on such data are posed in a drill-down manner, *i.e.*, they are first processed on coarse, highly compressed summaries corresponding to larger spatio-temporal volumes, and the approximate results obtained are used to focus on regions in the network that are most likely to contain result set data.

The standard wavelet compression algorithm used in DIMENSIONS assumes a grid topology in the data, and cannot directly handle an irregular placement of nodes. However, sensor network topologies are more likely to be irregular. To make the standard wavelet compression algorithm work with irregular topologies (without changing the wavelet algorithm itself), DIMENSIONS first convert an irregular topology data into a regular grid data before applying the standard wavelet compression algorithm. The regularizing procedure works as follows:

- Choose a coarse grid: choose a grid size that is coarser than the fine-grained data. In this case, 14x11 grid regions were chosen overlaying on the 140x110 grid acquired in step 1, and an average of 2 nodes per grid are expected.

- Averaging: For each such grid, average data from all nodes in the grid. Averaging data from multiple nodes will smooth the data and also reduce the noise in data. If there are N sensors in a certain square, the original noise per sensor is σ^2, and assuming the noise distribution at each sensor node is *i.i.d.*, then averaging the data obtains an aggregated measurement with variance σ^2/N (from the CLT theorem).

- Interpolation for empty grid cells: If there are no nodes in a certain cell, we use simple nearest neighbor interpolation to fill in these grids. Nearest neighbor interpolation fills the cell with data from the nearest non-empty cell. This can be easily implemented in a distributed setting: a cluster head that needs to receive data from 4 lower level nodes: and receives, say, only 3, fills in the empty grid with interpolated data. Finally, the resulting regular data is passed to the standard grid-based wavelet compression.

However, this data regularizing process may skew the original data and introduce error in the query processing. We use a simple example to illustrate why DIMENSIONS can be sensitive to topology.

Figure 5 shows an irregular topology. The solid black circles represent real nodes, each labeled with a sensor reading. To convert the topology into a grid, we overlay a grid on top of the original irregular topology. In general, when we overlay a grid on an irregular topology, one of the following scenarios will be observed: (1) multiple data will appear in an overlaid grid cell when the region is divided coarsely; (2) some grid cells will be empty when the region is divided in a very fine-grained manner; or (3) most likely, a combination of the above. In standard grid-based wavelet processing, at the base level of the data hierarchy, one data point is required from each grid. Thus we need to address the cases in which multiple data points are packed into one cell, or a cell is empty.

Using the regularizing approach described above, we establish a grid over the irregular topology in Figure 5, where the dotted circle represents the virtual node with values computed from the averaging or nearest neighbor interpolation procedure described above. In this data configuration, for a query about *the maximum value in the sensor field*, the wavelet operation will generate a reply *28* (assuming no compression error), but the real maximum sensor reading is *38*. The discrepancy is due to smoothing in the interpolation. Taking another example: for a query about *average sensor reading*, the wavelet operation on this will reply *14* (assuming no compression error), but the real average is *15*. The discrepancy is again due to interpolation and averaging when converting the original data into a grid.

In summary, the procedure to convert an irregular topology to grid data will introduce errors. Different irregular topologies will likely introduce different amounts of errors. Further, for a certain topology, different approaches used to construct a grid over the original data are also expected to deliver different results.

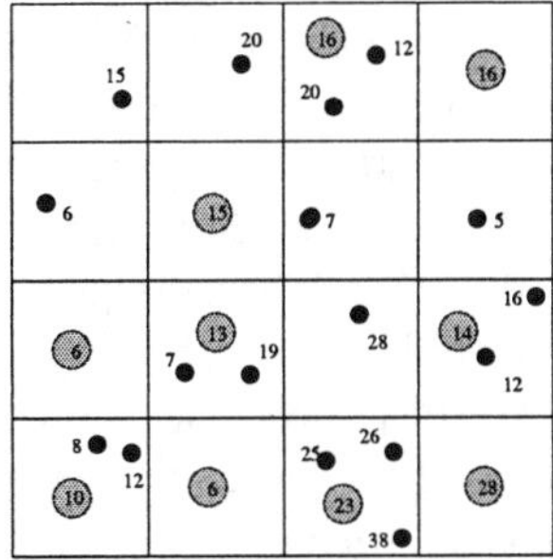

Figure 5: Irregular topology: converting an irregular topology to a grid will skew the data.

Next, we compare the algorithm performance under regular and irregular topologies. For regular topologies, we use an experimental grid data set; for irregular topologies, we use synthetic irregular data sets generated from the joint spatio-temporal model explained in section 3.2.

4.1 Evaluate DIMENSIONS using the grid data setDIMENSIONS [12] has been evaluated using a rainfall data set that provides 50km resolution daily precipitation data for the Pacific NorthWest from 1949-1994 [29]. The spatial setup comprises a 15x12 grid of nodes with 50-km spacing, each node recording daily precipitation values. While presumably this data set is at a significantly larger scale than what we would envision in a densely deployed sensor network, the data exhibits spatio-temporal correlations, providing a useful performance case study. A variety of queries can be posted on such a data set, such as range-sum queries, *e.g.*, total precipitation over a specified period from a single node or a region, or drill-down extreme value queries. In our comparison, we use maximum value queries as an example to demonstrate the algorithm performance. We present performance results for two queries: (a) GlobalDailyMaxQuery: which node receives max precipitation for a day in year X? (b) GlobalYearly-Max Query: which node receives the max precipitation for year X? The evaluation metric we used in this paper is mean square error. The error is defined as the difference between the measured query answer by dimensions, and the real answer as calculated by an optimal global algorithm with the same data input, normalized by the real answer. If we consider 15x12 grid data input as a coarse sample of the original phenomena, this mean square error can be deemed as the error in the system output compared with an approximate ground truth. Figures 6 and 7 show the mean square error vs. the level at which the drill-down query processing terminates.

In the DIMENSIONS hierarchy, each lower level stores twice the amount of data as the higher level. Therefore, as query processing proceeds down the DIMENSIONS hierarchy, consequently gaining access to more detailed information, as expected, the mean square error drops down gradually (shown in Figure 6 and 7). If the query accuracy followed the storage ratios (*i.e.*, 1:2:4:8:16), the accuracy improvement should be linear, however, it is usually super-linear, although the marginal improvement decreases with increasing levels. For instance, as shown in Figure 7, the marginal improvement in accuracy with just having the topmost level is approximately 66%, having the next level is 15%, followed by 6% and 2% respectively. Figure 6 exhibits similar results.

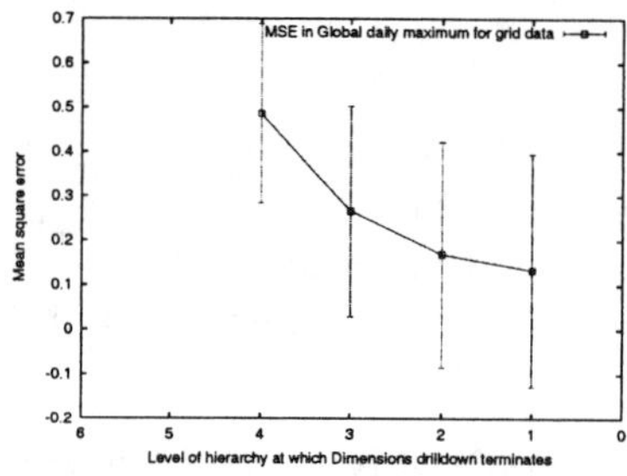

Figure 6: Error vs. Query termination level: Global daily maximum over original rainfall data.

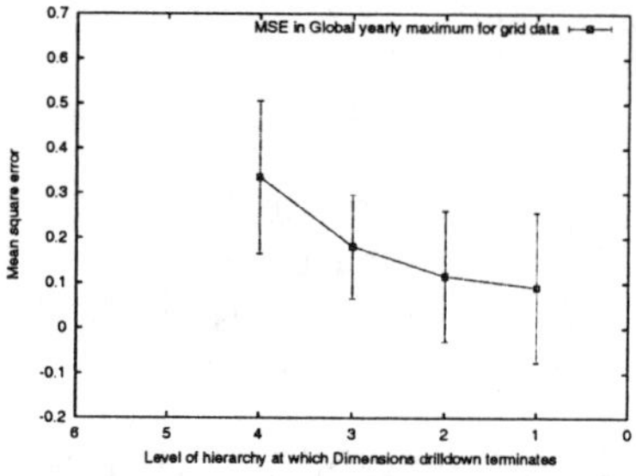

Figure 7: Error vs. Query termination level: Global yearly maximum over original rainfall data.

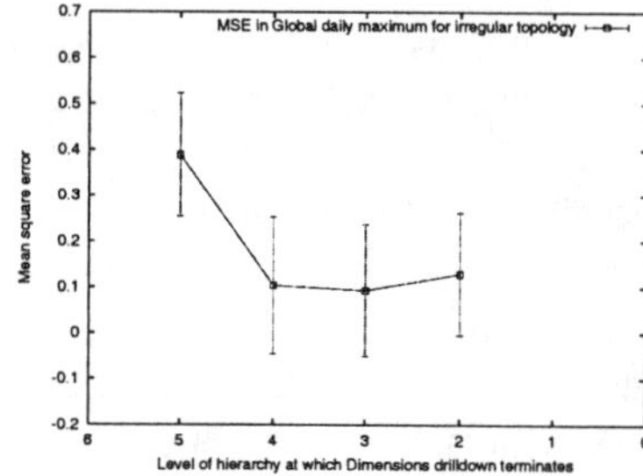

Figure 8: Error vs. Query termination level: Global daily maximum over irregular topology.

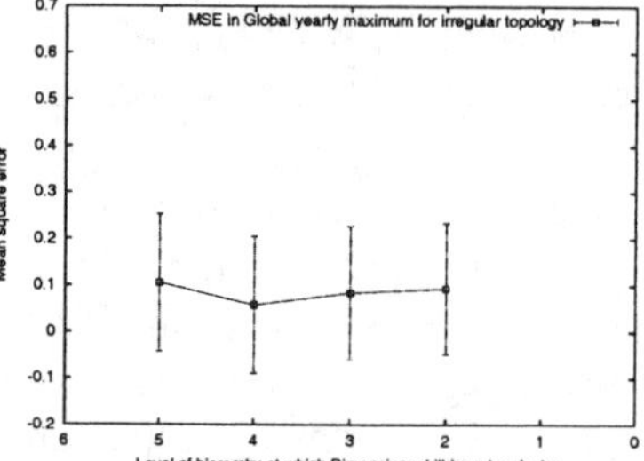

Figure 9: Error vs. Query termination level: Global yearly maximum over irregular topology.

4.2 Evaluating DIMENSIONS using the irregular data set. Here we evaluate the algorithm performance over an irregular topology. The precipitation data set we used to evaluate DIMENSIONS in section 4.1 is from a grid topology, thus it cannot be directly used here. On the other hand, the performance of the wavelet compression algorithm is sensitive to the correlation pattern in the data and the space of data correlation is too big to simulate exhaustively. Therefore, a randomly generated data set is considered unrealistic and of little value in the DIMENSIONS evaluation. Instead, we use a synthetic irregular topology derived from models of the precipitation data set we used in section 4.1. Following the procedure described in the beginning of section 4 , we generate a random topology consisting of 14x11x2 nodes from the precipitation data. The current implementation of the wavelet compression algorithm in the DIMENSIONS system works only with a regular grid topology. Thus, the regularizing step is used to convert the irregular data set into regular grid data.

4.2.1 Results and comparison

Due to the time consuming nature of wavelet operations, we illustrate our results with only a single topology. The results presented in this section are averaged for one irregular topology and multiple (*i.e.,* 44) queries.

Figure 8 and 9 show the results for the same queries as presented in section 4.1, GlobalDailyMax and GlobalYearlyMax, but for the irregular topology data. As in the regular grid case, we present the results in the form of the mean square error vs. the level at which the drill-down query processing terminates. The error here is the difference between the output of DIMENSIONS system with 14x11x2 irregularly placed nodes as input and the approximate ground truth. Here, the ground truth is approximated in the following way: Since the 14x11x2 irregularly placed nodes were selected from the 140x110 fine grained data, we first overlay a 14x11 grid over the 140x110 fine grained data, and for each grid, average 100 fine grained data points in the grid and obtain a value for each of 14x11 grid, and input this 14x11 data to the global algorithm. The ground truth was approximated by the output of the global algorithm.

As shown in Figure 8 and 9, DIMENSIONS behaves differently in an irregular setting from a regular one. In regular case (as shown in Figure 6 and 7) , we observe gradual error improvement with more levels of drill-down. However, in the irregular case (Figure 8 and 9), there is no consistent error improvement with more levels of drill down. For the topology that we studied, in some cases (Figure 8 and 9), the query error actually increases as more levels are drilled down. Such a behavior would not be expected in a regular grid topology. This may be due to effects caused by holes in topology and our corresponding interpolation procedure. An irregular topology might result in large regions where there is no data, and as described above, interpolation is used to fill in the empty cells in the regularization process. However, interpolation may smooth and therefore skew the data, which will introduce error in the query processing.

Note that if the graphs are averaged over multiple irregular topologies, some of the kinks in Figure 8 and 9 might be straightened out. However, we do not expect the same consistent improvement or the same order of improvement as with the regular grid data case.

Discussions The error that the user perceived and approximated by our computed mean square error is subject to multiple factors. We just name a few: the physical node topology, *i.e.,* the actual position of nodes; the data correlation statistics, *e.g.,* the form of the spatio-temporal model, and its interaction with wavelet compression; and the mechanism used to interpolate (in our case, nearest neighbor) in the regularizing procedure. If our objective is to study how irregular topology affects system performance, ideally we want to vary only the topology, and keep other parameters fixed. One way to achieve this is to model the data correlation in the experimental data set, and then use the same empirical model to generate regular and irregular topologies, which we will use to evaluate the algorithm's performance over regular and irregular data respectively. More specifically, we could use the output of the data modeling process as described in section 3 (*i.e.,* fine grained data set) as the empirical model. We sub-sample the same fine grained data set to obtain the regular and irregular sampled data as input to the DIMENSIONS system. The common model here is fine grained data. Since it is overly interpolated, it is deemed as an approximate model of the phenomena and used as the approximate ground truth to compute the mean square error. The error is thereby computed as: $\frac{QueryResponseOverDimesions - QueryResponseOverModel}{QueryResponseOverModel}$. We plan to explore this in our future work.

In summary, we used the synthetic data sets generated from modeling the spatio-temporal correlation in the experimental data set to evaluate DIMENSIONS over an irregular data set. DIMENSIONS in an irregular setting exhibited different behavior from the regular setting. This exposes the problem of DIMENSIONS's current regularization scheme. Thus our proposed synthetic data set helped to systematically evaluate the algorithm, and point out needed improvements.

In our case study, the precipitation data is from a regular grid topology. However, even if we have experimental data sets from irregular topologies, our proposed synthetic data generation approach will enable evaluating algorithms over different irregular topologies other than the particular setting used in the experimental data set. More importantly, it also allows us to evaluate algorithms over the same underlying data correlation model but different topology settings, so that we can study how the algorithm performance interacts with different parameters independently.

5 CONCLUSIONS

In this paper, we proposed to generate synthetic data from empirical models of experimental data, and use it to evaluate sensor network algorithms. In modeling the spatio-temporal correlation in the data, we draw heavily on spatial interpolation and geo-statistical techniques to implicitly or explicitly model the spatial correlation in the data, and generate irregular topology data that approximates the spatial correlation in the experimental data. To capture the correlation between the spatial and temporal domain, we use a joint space-time model inspired by [18]. We apply the proposed modeling techniques to a S-Pol radar data set. We propose spatial correlation (*i.e.*,variogram values) as a quantitative metric to evaluate synthetic data sets, which will be used to test sensor network protocols. Evaluation results show that most spatial interpolation techniques can closely approximate the spatial correlation and interesting properties (*e.g.*, edges) of the original data. We also use the DIMENSIONS system as a case study to show how synthetic data can help to evaluate the system over irregular topologies.

6 ACKNOWLEDGMENTS

The authors are grateful to NCAR for preparing and providing the S-Pol data set, especially Lynette Laffea and Bob Rilling for providing helpful explanations to our questions. We also wish to thank all LECS members for many helpful discussions, in particular, Hanbiao Wang, Vladimir Bychkovsky, Ben Greenstein for providing useful feedback on the draft, and Nithya Ramanathan, Jamie Burke, Tom Schoellhammer, and Adrienne Lavine for proofreading the draft. We are also grateful to Stefano Soatto for helpful feedback on interpolation algorithms. This work is made possible by a generous grant from the National Science Foundation through CENS.

REFERENCES

[1] The r project for statistical computing. In *http://www.R-project.org/.*

[2] Sandeep Bajaj, Lee Breslau, Deborah Estrin, Kevin Fall, Sally Floyd, Padma Haldar, Mark Handley, Ahmed Helmy, John Heidemann, Polly Huang, Satish Kumar, Steven McCanne, Reza Rejaie, Puneet Sharma, Kannan Varadhan, Ya Xu, Haobo Yu, and Daniel Zappala. Improving Simulation for Network Research. Technical Report 99-702b, University of Southern California, March 1999. revised September 1999, to appear in IEEE Computer.

[3] G. Box and G. Jenkins. *Time Series Analysis: Forecasting and Control* Holden-Day, 1976.

[4] J. Broch, D.A. Maltz, D.B. Johnson, Y.-C. Hu, and J. Jetcheva. A Performance Comparison of Multi-Hop Wireless Ad-Hoc Network Routing Protocols. In *Proceedings of the Fourth Annual ACM/IEEE International Conference on Mobile Computing and Networking (Mobicom'98)*, Dallas, TX, 1998.

[5] Ramon Caceres, Peter B. Danzig, Sugih Jamin, and Danny J. Mitzel. Characteristics of Wide-Area tcp/ip Conversations. In *ACM SIGCOMM,* 1991.

[6] Alberto Cerpa, Jeremy Elson, Deborah Estrin, Lewis Girod, Michael Hamilton, and Jerry Zhao. Habitat Monitoring: Application Driver for Wireless Communications Technology. *2001 ACM SIGCOMM Workshop on Data Communications in Latin America and the Caribbean*, April 2001.

[7] Alberto Cerpa and Deborah Estrin. Ascent: Adaptive Self-Configuring Sensor Networks Topologies. In *Infocom '02*, New York, June 2002.

[8] Samir R. Das, Charles E. Perkins, and Elizabeth M. Royer. Performance Comparison of Two On-Demand Routing Protocols for Ad-Hoc Networks. In *INFO*-COM, Israel, March 2000.

[9] Georg Dorffner. Neural Networks for Time Series Processing, *Neural Network World*, 6(4):447–468, 1996.

[10] Xin Li et al. New Edge-Directed Interpolation. *IEEE Trans. on Image Processing*, October 2001.

[11] P. Fryzlewicz, S. Van Bellegem, and R. von Sachs. A Wavelet-Based Model for Forecasting Non-Stationary Processes, *Submitted for publication.*, 2002.

[12] Deepak Ganesan, Deborah Estrin, and John Heidemann. Dimensions: Why do we Need a New Data Handling Architecture for Sensor Networks? In *Proceedings of the First Workshop on Hot Topics In Networks (HotNets-I)*, October 2002.

[13] Pierre Goovaerts. *Geostatistics for Natural Resources Evaluation*, Oxford University Press, Inc., 1997.

[14] Per Johansson, Tony Larsson, Nicklas Hedman, Bartosz Mielczarek, and Mikael Degermark. Scenario-Based Performance Analysis of Routing Protocols for Mobile Ad-Hoc Networks. In *Proc. ACM Mobicom*, Seattle, WA, 1999.

[15] Almudena Konrad, Ben Y. Zhao, Anthony D. Joseph, and Reiner Ludwig. A Markov-Based Channel Model Algorithm for Wireless Networks, in *Proceedings of Fourth ACM International Workshop on Modeling, Analysis and Simulation of Wireless and Mobile Systems, ACM MSWiM*, July 2001.

[16] D.G. Krige. Two-Dimensional Weighted Moving Average Trend Surfaces for Orevaluation. In *Journal of the South Africa Institute of Mining and Metallurgy,* Vol. 66, 1966.

[17] P.C. Kyriakidis and A.G. Journel. Geostatistical Space-Time Models. in *Mathematical Geology,* Vol. 31, 1999.

[18] P.C. Kyriakidis, N.L. Miller, and J. Kim. A Spatial Time Series Framework for Modeling Daily Precipitation at Regional Scales. In *82nd Annual Meeting of the American Meteorological Society*, January 2002.

[19] Y.Ogata. Space-Time Point-Process Models for Earthquake Occurences. In Ann. *Inst. Statist. Math.*, Vol. 50, 1998.

[20] S. Park, A. Savvides, and M.B. Srivastava. Sensorsim: A Simulation Framework for Sensor Networks. In *MSWiM*. ACM, August 2000.

[21] S. Park, A. Savvides, and M.B. Srivastava. Simulating Networks of Wireless sensors. In *2001 Winter Simulation Conference*. ACM, 2001.

[22] N. Abu-Ghazaleh S. Tilak and W. Heinzelman.Infrastructure Tradeoffs for Sensor networks. In *ACM 1st International Workshop on Sensor Networks and Applications (WSNA '02)*. ACM, September 2002.

[23] F. P. Schoenberg. Consistent Parametric Estimation of the Intensity of a Spatial-Temporal Point Process. In *Ann. Inst. Stat. Math. (in review)*. Wiley, NY.

[24] F.P. Schoenberg, D.R. Brillinger, and P.M. Guttorp. Point Processes, Spatial-Temporal. In *Encyclopedia of Environmetrics*, Vol.3. Wiley, NY.

[25] C. Schurgers, V. Tsiatsis, and M. Srivastava. Stem: Topology Management for Energy Efficient Sensor Networks. In *IEEE Aerospace Conference,* Big Sky, MT, March 2002.

[26] Y. Theodoridis and M. Nascimento. Generating Spatiotemporal Data Sets in the WWW. In *SIGMOD Record,* Vol. 29 (3), 2000.

[27] W. Du Mouchel, C. Volinsky, T. Johnson, C. Cortes, and D. Pegibon. Squashing Flat Files Fatter. In *Proc. KDD*, 1999.

[28] R. Webster and M. Oliver. *Geostatistics for Environmental Scientists*, John Wiley & Sons, Inc., 2001.

[29] M. Widmann and C.Bretherton. 50 km Resolution Daily Preciptation for the Pacific Northwest, 1949-94, http://tao.atmos.washington.edu/data_sets/widmann.

[30] Y. Xu, J. Heidemann, and D. Estrin. Geography Informed Energy Conservation for Ad-Hoc Routing. In *ACM MOBICOM'01*, Rome, Italy, July 2001.

[31] Y. Yu, D. Ganesan, L. Girod, D. Estrin, and R. Govindan. Modeling and Synthetic Data Generation for Fine-Grained Networked Sensing. *UCLA/CENS Tech. Reports 15*, 2003.

[32] X. Zeng, R. Bagrodia, and M. Gerla. Glomosim: a Library for Parallel Simulation of Large-Scale Wireless Networks. In *Proceedings 12th Workshop on Parallel and Distributed Simulations – PADS '98*, May 1998.

Appendix 1: Brief introduction of variogram

A variogram is used to characterize the spatial correlation in the data. The variogram (also called semivariance) of a pair of points x_i and x_j is defined as $\gamma(x_i, x_j) = \frac{1}{2}\{Z(x_i) - Z(x_j)\}^2$. We can also define semivariance as a function of lag (*i.e.*, the separation between two points, is distance in one dimension, or a vector with both distance and direction in two and three dimensions), h:

$$\gamma(h) = \frac{1}{2}E[\{Z(x) - Z(x+h)\}^2] \tag{4}$$

For a set of samples, $z(x_i)$, i=1, 2, ..., $\gamma(h)$ can be estimated by

$$\hat{\gamma}(h) = \frac{1}{2m(h)} \sum_{i=1}^{m(h)} \{z(x_i) - z(x_i + h)\}^2 \tag{5}$$

where $m(h)$ is the number of samples separated by the lag distance h.

Data in high dimensions might add complexity in modeling variograms. If data lie in a high dimensional space, variograms are first computed in different directions separately. If variagrams in different directions turn out to be more or less the same, the data are isotropic, then sample variograms can be averaged together. Otherwise, data in different directions need to be modeled separately.

Energy Efficient Channel Allocation in Publish/Subscribe GeoSensor Networks

Saravanan Balasubramanian and Demet Aksoy

Department of Computer Science
University of California, Davis
{balasubr, aksoy}@cs.ucdavis.edu

ABSTRACT

Recent applications in sensor networks contain advanced methods for geo, air, and water monitoring. As the number of sensors being deployed is increasing as well as the computation power of these sensors, there is an emerging need to exchange mass amounts of data to analyze observations made. We study efficient and interference-aware communication mechanisms between geosensors. In this paper, we present an online broadcast scheduling algorithm for a publish/subscribe sensor network. Using a simple subscription protocol, sensors publish their topic information to its neighbors who can then subscribe to interested topics. Our online scheduling protocol creates a schedule without any initial global knowledge, attempts to reduce the number of channels in the schedule and the amount of transmissions made by a sensor. Analysis of our simulation results show that our online algorithm is competitive to the centralized vertex-coloring algorithm while decreasing total energy consumption by the sensors.

1. INTRODUCTION

Recent advances in wireless technology and the development of small, low-cost, low-energy electronics has led to increased research in the area of wireless sensor networks. A wireless sensor network is a highly distributed system consisting of small, wireless, low-energy, unattended sensors. There are various defense, scientific, and engineering applications that use wireless sensor networks -- military reconnaissance and surveillance, environmental and habitat monitoring, and wildlife tracking.

A sensor's basic functionality is to acquire data from the physical environment, process the data, and communicate the information to other sensors or users. The functionalities of a sensor have many similarities to the publish/subscribe paradigm that uses push technology [CAR01]. The operations in this environment are simple. Publishers advertise the information they are providing, and clients subscribe to the information if interested. When the publisher senses an event, it disseminates the generated data to its subscribed clients. This allows the publishers of continually changing information to reach large numbers of clients more efficiently. Publish/subscribe technology is deployed widely to relay real-time information such as stock quotes, weather information, and news reports to subscribed clients.

To define and illustrate the problem, we present a sensor network for environmental habitat studies. In this network, each sensor monitors a single factor affecting the habitat. This factor could be air temperature, humidity, barometric pressure, wind speed, UV intensity, solar radiation etc. Many scientific calculations in environmental sciences involve a combination of these factors. We provide a few examples below:

1. The wind chill factor [WEB03] is a function of air temperature and wind speed.
2. Heat stress [WEB03] measures the apparent temperature and is a function affected by air temperature and humidity.
3. Evapotranspiration [WEB03] measures the amount of water lost from the ground due to evaporation or transpiration. It is a function of air temperature, humidity, wind speed, pressure, solar radiation, and time.

The above examples show that there can be situations where a sensor requires data from other sensors either to confirm an event or to generate new data. Assume there are two sensors, A and B, generating temperature and humidity data respectively. When an event occurs, Sensor B disseminates the generated data to its clients. Upon receiving the humidity data from Sensor B, Sensor A can combine it with its own data to calculate heat stress, which the sensor transmits to other subscribed clients. Such communication can take place using random access models. However, without coordination, data dissemination could result in many collisions.

This paper focuses on avoiding multi-user interference, which occurs when several sensors simultaneously communicate within each other's transmission range. In an uncoordinated environment, the signals will collide garbling the message and requiring the sensors to retransmit. If the sensors are not coordinated, collisions result in significant energy consumption due to retransmissions. Sensors are typically energy-limited, and therefore should limit unnecessary retransmissions, which means avoiding collisions is of extreme importance.

There have been a number of studies for developing offline schedules to avoid collisions [RAM92][RAM99][JJU99][BJO03]. Creating an offline schedule for a sensor network requires perfect knowledge of the network topology, and the publishers and subscribers in the network. Obtaining perfect a-priori knowledge is not practical in situations where scientists and researchers deploy low-cost, micro sensors in great quantity throughout a habitat. Such an occurrence would create a dynamic network topology that requires sensors to discover their neighbors and compute the schedule online. Another advantage of an online scheduling algorithm is that it allows the network to add and remove sensors after the initial deployment and sensors to change their subscription pattern, i.e., when a need for further analysis is required.

In this study, we introduce our on-line scheduling effort to guarantee collision-free sensor communication with an objective of reducing energy consumption for sensor communication. This paper provides a framework that allows sensors to create a schedule that avoids collisions, does not require any initial global knowledge, reduces the number of channels in the schedule, and makes use of spatial features.

The paper is organized as follows. We first discuss issues related to link and broadcast scheduling in sensor networks in Section 2. In Section 3, we present our online algorithm and explain its operations. Section 4 explains the simulation setup, metrics, and discusses the results. Section 5 describes related works. Section 6 concludes the paper and provides future directions in the area of scheduling.

2. APPLICATION-AWARE SCHEDULING

Application-aware scheduling involves multiplexing the medium into channels and assigning them for specific communications between sensors. Sensors can transmit only in their assigned channels. A transmission schedule is constructed and deployed in the network to create a collision-free environment. The challenge lies in the scheduling, which has to avoid collisions and reduce the number of channels in the schedule.

There exist two types of scheduling: link and broadcast scheduling [RAM92][RAM99]. In link scheduling, an algorithm assigns each pair-wise communication to a channel. Two links may not be assigned to the same channel if they are adjacent to each other or if a third link exists from the sender of one link to the receiver of the other link. A broadcast scheduling algorithm, on the other hand, constructs a schedule such that two sensors may not be assigned to the same channel if they are adjacent to each other or if they have a common neighbor. This will ensure that neighbors receive a transmission by a sensor without collisions. In this paper, we focus on broadcasting scheduling.

There have been a number of studies for developing offline schedules for this problem [RAM92][RAM99][JJU99][BJO03]. Our focus is on an on-line schedule that can support a dynamic network topology that evolves according to the requirements of the sensor network.

Our scheduling objective is to guarantee a collision free publication environment regardless of the cause of interference. Collisions can be categorized into two: 1) those that are due to primary and 2) those that are due to secondary interference [RAM92]. A primary interference occurs when a sensor has to perform two activities in the same channel. For example, a sensor cannot receive transmissions simultaneously from multiple sensors or perform a transmission while it receives. A secondary interference occurs when a sensor receives multiple transmissions because it is in the range of another transmission.

3. OUR ONLINE APPROACH

The goal of our online approach is to create a schedule that avoids collisions, reduces the number of channels in the schedule, and makes use of spatial features without having any initial knowledge about the topology. To achieve this, the broadcast medium through which communication takes place should use multiple channels. We reuse channels as long as there is sufficient spatial separation between sensors. For instance, two sensors can transmit in the same channel at the same frequency if the reuse distance separates them. Thus, the spatial and temporal natures of the wireless links generate great challenges in developing an optimal schedule. Based on this intuition we develop the framework presented below.

3.1 Messaging Framework

The sensors exchange a series of messages that allow them to discover their neighbors, advertise their topics, subscribe to other sensors, and reserve channels for transmission and reception.

Upon sensor deployment and activation, each sensor must advertise (by sending an ADV message) its unique identification number and the publication topic. Note that only those who are in the coverage area can receive and reply to ADV messages. When a sensor hears an advertisement of interest, it replies with an interested message (INT). Sensors not interested in the published topic can either send a message saying they are not interested (NOT INT) or ignore the message according to their battery power. [1] If a sensor chooses to send a NOT INT message, it reveals additional information about its existence which is used in later steps of channel allocation optimizations.

When a sensor transmits a message, other sensors in the coverage area overhear the message. Based on the messages overheard, publishers and subscribers discover sufficient information to map their local topology. Overall, we can represent the data flow in the network using a directed graph. The vertexes in the directed graph represent the sensors and the edges represent the data dissemination flows from a publisher to its subscriber.

---

[1] To unsubscribe from a previously subscribed topic, a sensor transmits a NOT INT message.

Figure 1 shows a simple example. In the figure, we illustrate a grid topology where sensors are located equidistantly from each other. The transmission range is equal to the horizontal and vertical distance between each sensor. In this example, Sensor B advertises its publication information. Sensors within the coverage area (A, C, and Y) hear this advertisement. Only Sensor A is interested in Sensor B's data and replies the ADV message with an INT stating its interest. During this process, Sensor X overhears this communication and saves this information to learn about the sensors in the vicinity. In Figure 1b we draw a directed edge from Sensor B to Sensor A to indicate the publish/subscribe relation of the two sensors. Since there is no communication with Sensor C and Y (both replied with a NOT INT message), we do not draw directed edges (Figure 1c).

Figure 1d presents the final graph presentation when all nodes complete their message exchange. Note that this overall graph presentation is what can be inferred by topology information of individual sensors. Each sensor has sufficient knowledge to pursue dynamic channel allocation according to neighbor subscriptions. In this example, all sensors are publishers but not all are subscribers. We see that Sensor C is not interested in any publication. Thus, there is no incoming edge for Sensor C.

During the advertisement and registration process, each sensor also records messages sent/received by its neighbor to use their subscription information for future use.

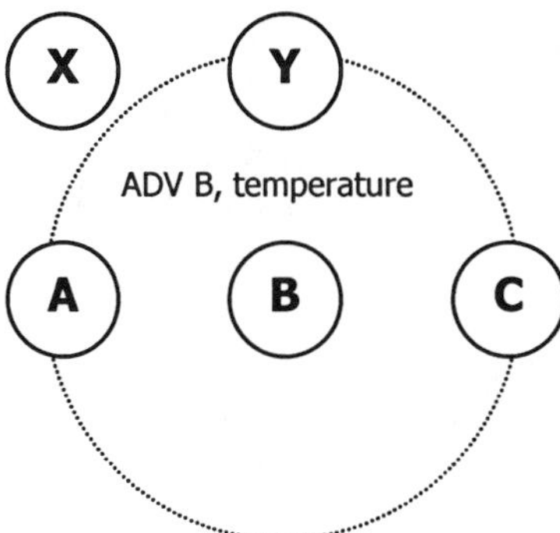

Figure 1a: Figure 1a to Figure 1d illustrates how our algorithm forms a directed graph through advertisement and subscription. Initially, all sensors advertise their existence and topic to its neighbors. In the above example, Sensor B advertises to its neighbors.

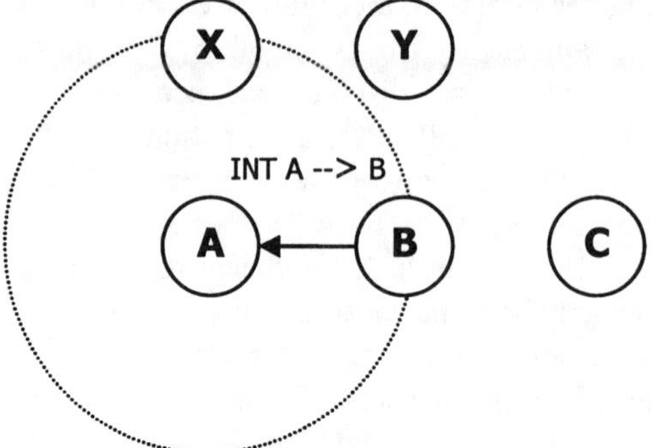

Figure 1b: Sensor A is interested in Sensor B's data, and therefore subscribes to its data.

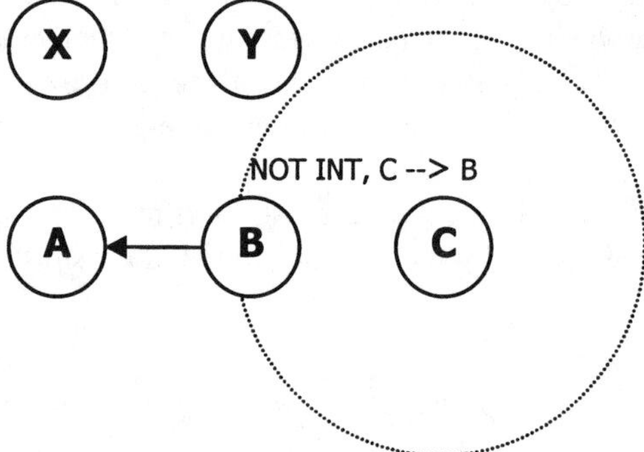

Figure 1c: Sensor C is not interested in Sensor B's data. In this example Sensor C chooses to send a NOT INT message. Now sensor B learns about neighbor C.

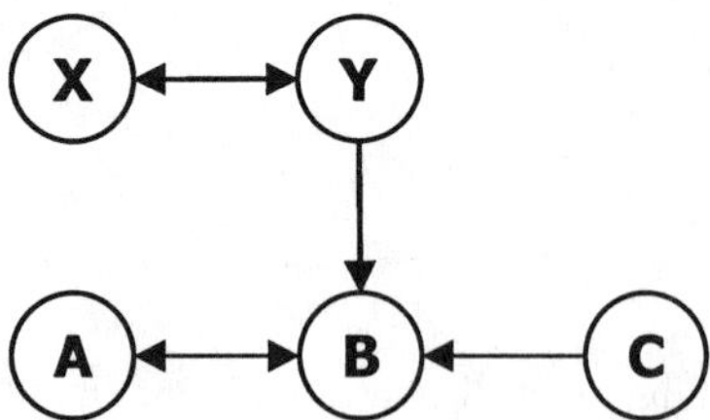

Figure 1d: Eventually, we can present all subscriptions using a directed graph after all sensors exchange messages. Each vertex represents a sensor and an edge represents the data dissemination flow from publisher to subscriber.

Based on the resulting directed graph, each sensor determines its schedule through a simple reservation protocol. When a sensor wants to reserve a channel for transmission, it sends a request (REQ) message specifying the channel to reserve at its receivers. Only sensors publishing data can request channel reservations.

Upon receiving the REQ message, if the requested channel is free, then the sensor responds with an acknowledgement (ACK) message.[2] If the receiver does not have a free channel at the specified channel number, then it responds with a negative acknowledgement (NACK) message. The sender of the NACK message also includes its current schedule, which allows the requesting sensor to determine the next available free channel in a shorter amount of time. Note that the sensors make the reservations on a first-come first-served basis. A sensor might construct a different reservation schedule if the order in which the sensors obtain access to the medium changes. Unlike a CTS message of IEEE 802.11 that immediately follows an RTS message, the ACK or NACK message of our algorithm may not immediately follow the REQ message. However, we use application layer message exchanges without causing any problems.

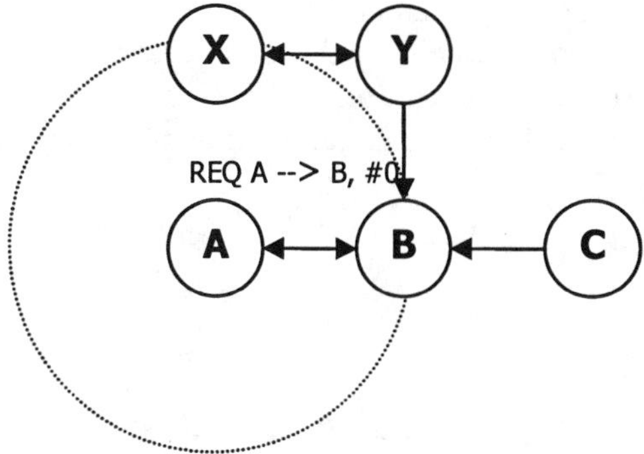

Figure 2a: Sensor A sends a request to B to transmit on channel 0.

[2] Similar to the advertisement phase, when a sensor makes a request, all neighboring sensors within its range overhear the REQ and ACK messages and make note of the reservation in order to reduce the number of future messages need to be exchanged when an update occurs in the network.

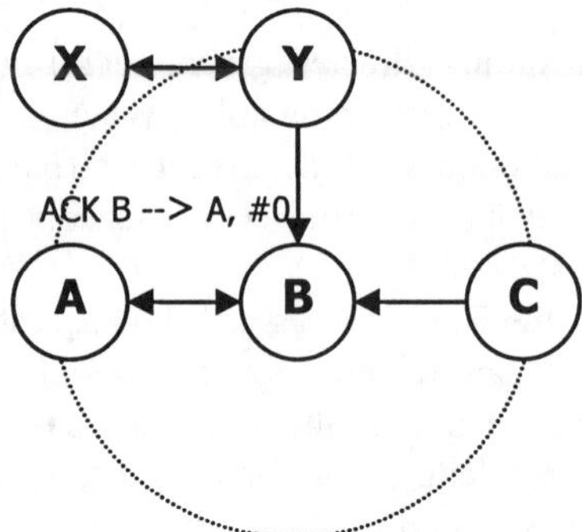

Figure 2b: Sensor B sees that its channel is free, and sends an ACK message to Sensor A.

Figure 2 illustrates the reservation process. In Figure 2a, Sensor A tries to allocate the first unused channel, namely channel number 0, for the publications it will transmit to Sensor B. When Sensor B replies with an ACK message, Sensor C overhears the acknowledgement from Sensor B; so it concludes that a neighbor will be listening to publications on channel number 0. Trying to reuse this channel will be conflicting for messaging due to primary interference. Therefore without exchanging additional messages, it will try to reserve channel 1 (not 0) when trying to allocate the publication channel for Sensor B. In this example, when sensor B receives the allocation request for channel 1, there is no conflict for this channel. Therefore no further messaging is required.

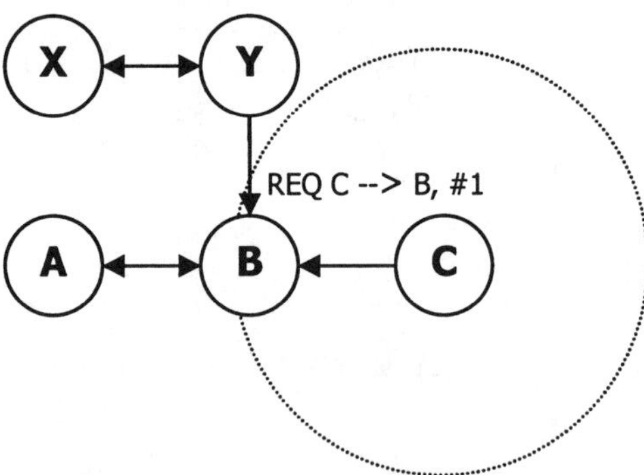

Figure 2c: - Since Sensor C also received the ACK message (Figure 2b), it knows that channel 0 is reserved by Sensor B. Therefore, it sends a request to transmit on channel 1 to Sensor B.

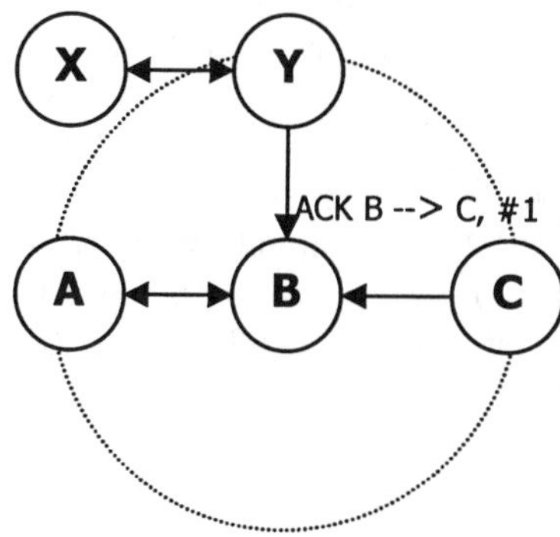

Figure 2d: An ACK is sent to Sensor C for the request made in Figure 2c. If for some reason Sensor B had channel 1 reserved, then Sensor B would return a NACK message to Sensor C. In the NACK message, Sensor B's current schedule will also be transmitted.

3.2 Channel Status Indicators

By listening to other sensors' REQ and ACK messages, a sensor can learn about its neighbors' schedule and available channels. Our algorithm uses status indicators to determine a channel's status. We use the following status indicators: FREE (F), TRANSMIT (T), RECEIVE (R), and PARTIALLY_FREE (PF). The FREE status indicates that the sensor could use the channel to either transmit or receive. The TRANSMIT and RECEIVE statuses are set whenever a sensor transmits or receives on that channel. The PARTIALLY_FREE status is set when a sensor hears a REQ or ACK message that indicates a reservation on a channel. However due to the spatial properties, the same channel can still be used without causing any interference.

For example in Figure 2a when Sensor A transmits a REQ message, Sensor X updates its status at channel 0 with a PF status. This indicates that Sensor X can still transmit or receive from a restricted set of neighbors on the same channel. To avoid collisions, Sensor X can only transmit if its receivers are not neighbors of Sensor A, and its neighbors are not receivers of Sensor A. Similarly, when a sensor hears an ACK message, it updates its channel with the PF status indicator. In Figure 2b, when Sensor Y hears the ACK sent by Sensor B, it updates its status at channel 0 to FR. This indicates that Sensor Y cannot transmit on channel 0 but can receive from another neighbor at that channel number provided that no collision can happen. Note that after receiving this information, Sensor Y cannot transmit on this channel because a transmission would cause a collision at Sensor B.

Using status updates allows neighbors to learn about each other's schedule and prevents unnecessary communication from occurring, which helps in reducing energy consumption.

3.3 Channel Allocation Characteristics

As described previously, an important aspect of our reservation protocol is that it makes use of spatial features. As stated previously, the broadcast channel through which communication occurs can be reused as long as there is sufficient spatial separation. Two sensors can confidently transmit in the same channel at the same frequency if the reuse distance separates them. In Figure 3, we illustrate the spatial feature where Sensors X and C transmit simultaneously using the same channel. There is no collision between the transmissions.

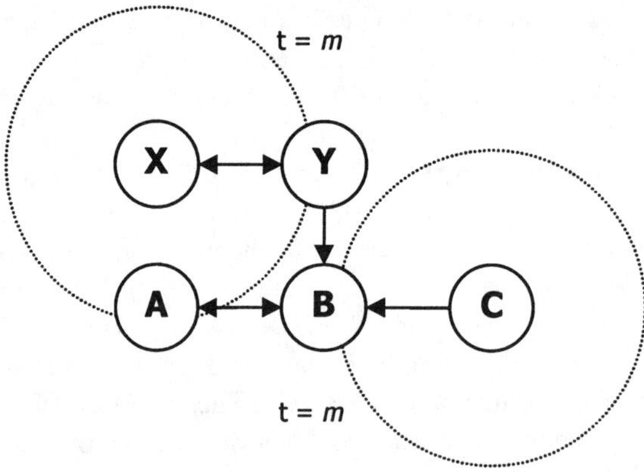

Figure 3: Example of spatial reuse. Sensor X and C can transmit in channel *m* without having any collisions.

Figure 4 illustrates a stronger spatial feature. While Sensor A transmits to Sensor B on channel *m*, Sensor X can simultaneously transmit to Sensor Y. Their transmission signals do collide but do not interfere at their respective receivers.

In our algorithm, we aim at allowing such channel reuse to the extent possible. Therefore listening to ongoing communication and transmitting neighbor information along with the request message is an important factor in the system.

An important focus of our work is to consider energy consumption at the sensors. Since sensors are typically energy-limited, it is ideal to reduce the number of transmissions made in the network. Using the topology in Figure 4 as an example, Sensor Y has sensors X and B subscribed to it. In this situation, it is ideal to transmit to both X and B in the same channel in order to save the transmission power of Sensor Y. It is possible that the channel will be free in Sensor X but not in Sensor B such that Sensor X will send an ACK message while Sensor B will communicate a NACK message. When this occurs, our algorithm chooses to make a new reservation for both sensors using another channel. Reserving channels in such a manner reduces the number of transmissions made by the publisher to its subscribers.

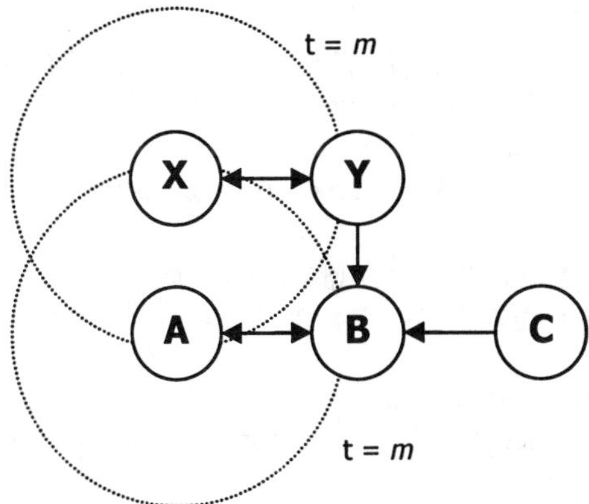

Figure 4: Example of another spatial reuse feature. Sensors A and X can simultaneously transmit to Sensors B and Y respectively because their transmission signals do not collide at their respective receivers.

4. PERFORMANCE EVALUATION

To evaluate the protocol and algorithm, we wrote a Java simulator that creates sensors, assigns each sensor's neighbors, and builds an edge between sensors that subscribe to topics of interest.

To evaluate the performance of our algorithm, we compare our algorithm to an offline vertex-coloring algorithm [RAM92]. The vertex-coloring algorithm starts with perfect knowledge of the topology. Based on this topology, it chooses a vertex and colors all outgoing edges. It continues this process until all vertices have been colored.

In the case of our online algorithm, sensors have no initial knowledge of the network topology. Once the sensors have determined their subscribers, the scheduling begins through message exchanges. Each sensor has a queue that contains messages that need to be transmitted. When a sensor obtains the medium, it transmits the first message in its queue. If there are no messages in the queue, then it checks to see if it has any transmissions to subscribers that need to be scheduled. If it does, then it sends a request to its subscribers. Otherwise, it only listens to the ongoing communication for future reference.

In our simulations, we implemented two types of grid topology. In the first grid network, a sensor has at least 2 and at most 4 neighbors. The second grid topology has a range of 3 to 8 neighbors per sensor. Subscriptions are determined based on connectivity rate. The connectivity rate dictates the probability of a sensor subscribing to its neighbor.

4.1 Metrics

We measure the performance of the algorithm using the following metrics.

- *Largest schedule length.* For networks based on TDM (Time Division Multiplexing) the schedule length determines the time interval between successive transmissions. In this respect we measure the largest schedule length as

  $$\text{Largest Schedule Length} = \max(\text{schedule length of } n_i),\ 1 = i = N$$

  where N is the total number of sensors and n_i is the ith sensor within the topology.
- *Average deviation from the largest schedule length.* In order to reflect the variation of schedule length of each sensor, we also measure the deviation from the largest schedule.
- *Energy inefficient channel usage (EICU).* Ideally, a sensor should transmit to all its subscribers using a single channel. A sensor uses more energy when it transmits to its subscribers using more than one channel. Energy Inefficient Channel Usage metric measures this inefficiency as:

  $$EICU = (\Sigma\,(CU_i - ICU_i)) / N$$

  where N is the number of sensors in the network, CU is the actual channel usage to receive from/transmit to neighboring sensors based on the resulting schedule, and ICU is the ideal channel usage based on the directed graph that represents the topology, i.e.,

  $$(ICU_i) = x_i + y_i$$

  where x_i is the minimum number of publications that must be performed by sensor i (0 if it contains zero outgoing edges, 1 otherwise - a single transmission for all subscribers), and y_i is the number of incoming edges (one channel per each transmission to avoid primary interference). Thus, EICU is the average difference between the non-free channels and the ideal channel usage.
- *Average number of messages exchanged.* Average number of message exchanged = (Σ number of messages sent by Sensor_i) / N

4.2 Experiments

In the first experiment, we used a grid topology with a maximum of 8 neighbors. We use a pair of connectivity rates: 5% and 95%. In the 95%

connectivity, almost every sensor subscribes and publishes to a neighboring sensor.

We first run the subscription protocol and make channelization decisions. At the end of the run, we report the maximum schedule length obtained by our online algorithm and the previously proposed centralized algorithm.

Figure 5 plots the maximum schedule length as the number of sensors in this topology increase from 1 to 4900 for a topology of low connectivity, i.e., where most of the sensors in the network are not interested in publications of the ir neighbors. Figure 5 shows that the number of sensors in the network does not greatly affect the schedule length. When the network size increases beyond 1000 sensors, the schedule length only increases at a very low rate. As can be seen in the figure, our online algorithm performs close to the centralized algorithm.

In Figure 6 we plot the results when we repeat the experiment for a topology of high connectivity of 95%. At this connectivity rate, almost each sensor has a publish/subscribe relation with a neighbor. As can be seen in the figure, our online algorithm performs close to the centralized algorithm for both connectivity rates. For the most part, our online algorithm produced schedule lengths equal to or one greater than the centralized algorithm. This is a quite reasonable penalty for the energy-conserving feature of our online algorithm.

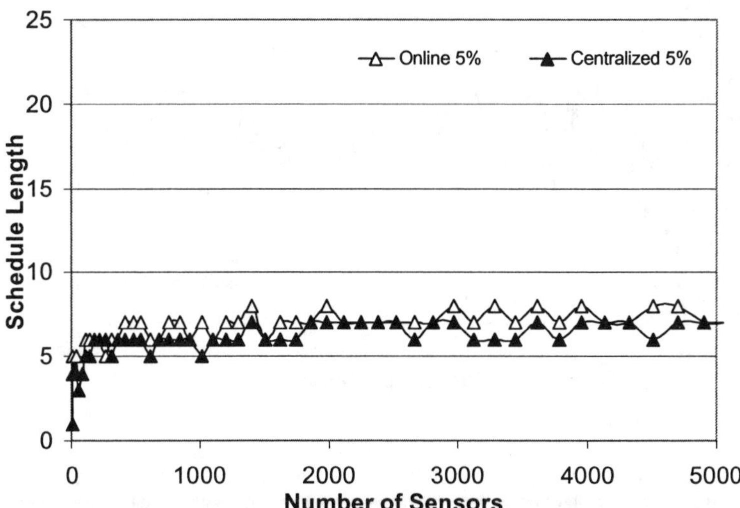

Figure 5: Low Connectivity Rate: In the case where most sensors are not subscribers to neighboring sensors we observe the largest schedule length between the centralized and online algorithms to be very close.

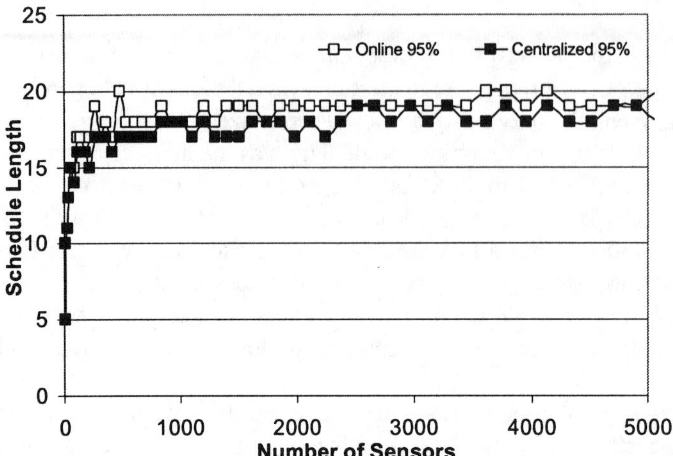

Figure 6: High Connectivity Rate: In this case most sensors subscribe to neighboring sensors. Even in case the difference is not significant.

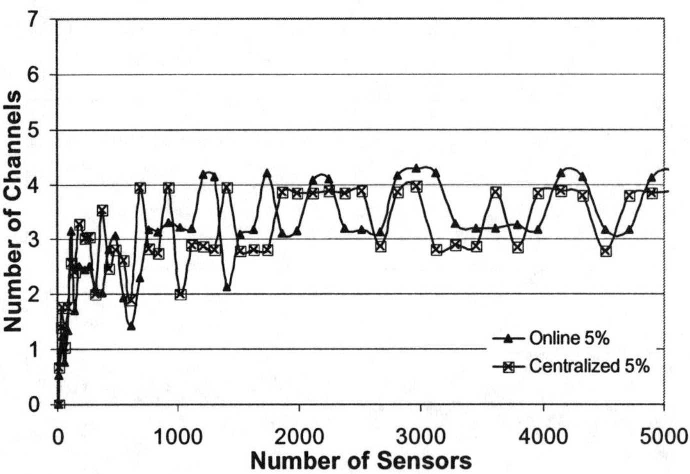

Figure 7: In the case where most sensors are not subscribers to neighboring sensors, we observe the standard deviation for individual sensors to be almost equivalent to the schedule length.

We plot the average deviation of schedule length obtained by individual sensors for the same experiment in Figures 7 and 8 for different connectivity rates. In the figure, we see that the average deviation increases as the network size increases. However, the deviation within the whole range is limited around 3 to 4 channels.

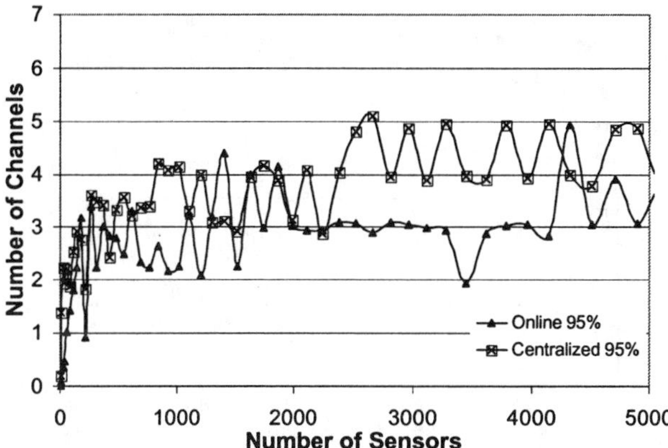

Figure 8: In the case where almost all sensors are subscribers to neighboring sensors, on the average we observe a lower standard deviation.

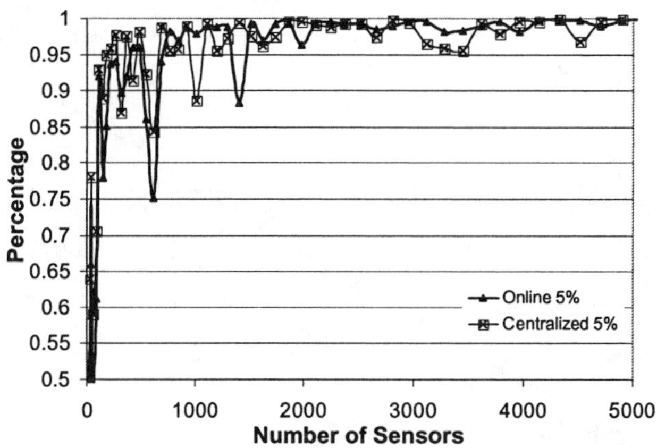

Figure 9: In the case where most sensors are not subscribers to neighboring sensors, we observe a little difference when we consider sensors that have a different schedule length.

Figures 9 and 10 plot the percentage of sensors that have a smaller schedule length than the maximum length observed during the experiment. The figures show that almost the majority of the sensors (99%) have a different schedule length than the largest schedule length regardless of the connectivity rate. In these experiments, in terms of deviation from maximum schedule length, there was not much difference between the centralized and our online algorithm.

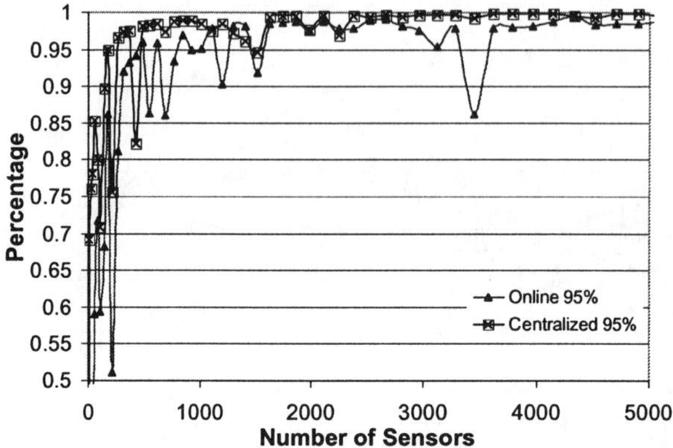

Figure 10: For high connectivity rate, the difference between the online and the centralized algorithm becomes a bit more pronounced.

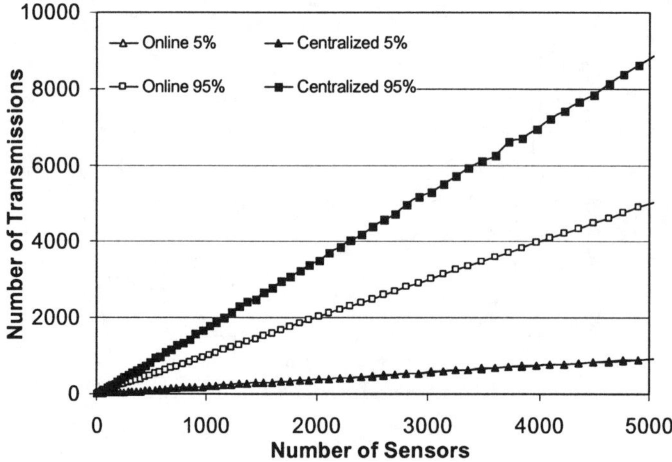

Figure 11: Total number of transmissions for centralized and online algorithm at varying connectivity rates.

In order to evaluate the cost of our heuristics, we plot the total number of transmissions made during the same experiment in Figure 11. As seen in Figure 11, the total number of transmissions made by the centralized algorithm is much larger than the online algorithm. For instance, for 95% connectivity it is more than twice. The difference between the number of transmissions made by the centralized and the online algorithms increase as the connectivity rate increases. The advantage of reserving all subscribers in the same channel is that it reduces the number of transmissions made by the publisher. This translates to much better energy savings than the centralized algorithm especially over a long period of time.

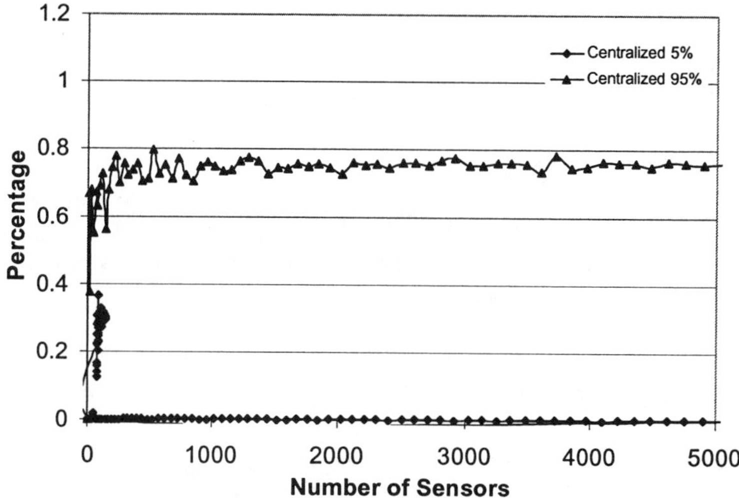

Figure 12: Average EICU for the centralized algorithm at varying connectivity rates. The results are for grids with 2-4 neighbors.

Figures 12 and 13 plot the EICU for the centralized algorithm with varying connectivity rates. In Figure 12, the topology is based on 2-4 neighbors, and in Figure 13 it is based on 3-8 neighbors. Note that we do not plot our online algorithm in these graphs. Our algorithm consistently produces a constant 0% EICU in both cases and therefore has been omitted from the graphs. In the figure, we see that as the connectivity rate increases the EICU of the centralized algorithm increases because there is an increase in the number of reservations that need to be made. For both connectivity rates, our on-line algorithm performs orders of magnitude better than the centralized algorithm. This suggests that we are able to create a power-efficient schedule in a much shorter time compared to the centralized algorithm.

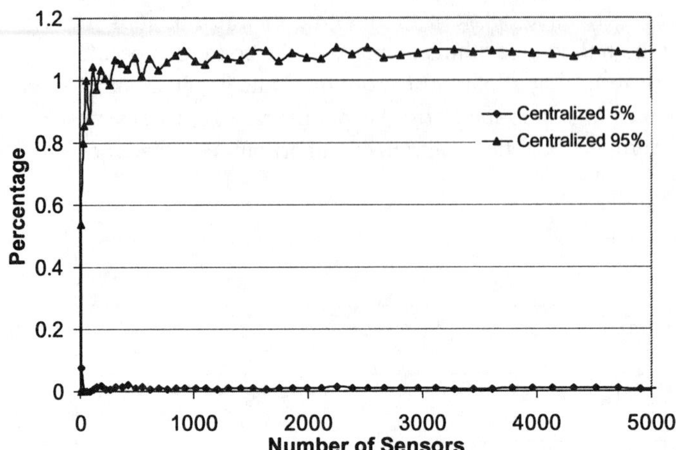

Figure 13: Average EICU for the centralized algorithm at varying connectivity rates. The results are for grids with 3-8 neighbors.

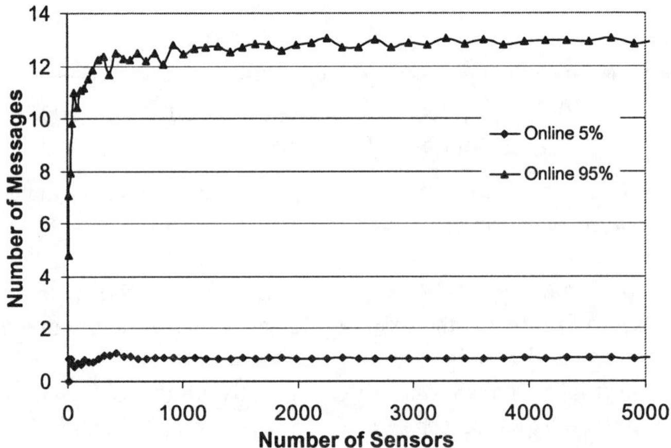

Figure 14: Average number of control message exchanges for a network topology containing 2-4 neighbors.

In order to evaluate the cost of our heuristics, we plot the total number of transmissions made during the same experiment in Figure 14. As seen in Figure 14, the total number of transmissions made by the centralized algorithm is much larger than the online algorithm. For instance, for 95% connectivity it is more than twice. The difference between the number of transmissions made by the centralized and the online algorithms increase as the connectivity rate increases. The advantage of reserving all subscribers in the same channel is that it reduces the number of transmissions made by the publisher. This translates to much better energy savings than the centralized algorithm over a long period.

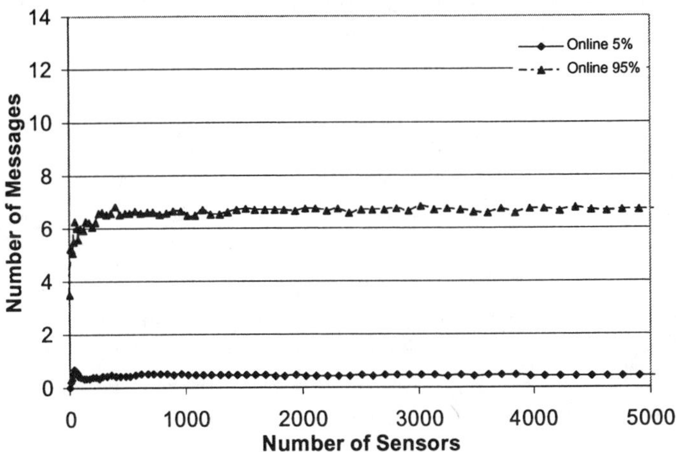

Figure 15: Average number of control message exchanges for a network topology containing 3-8 neighbors.

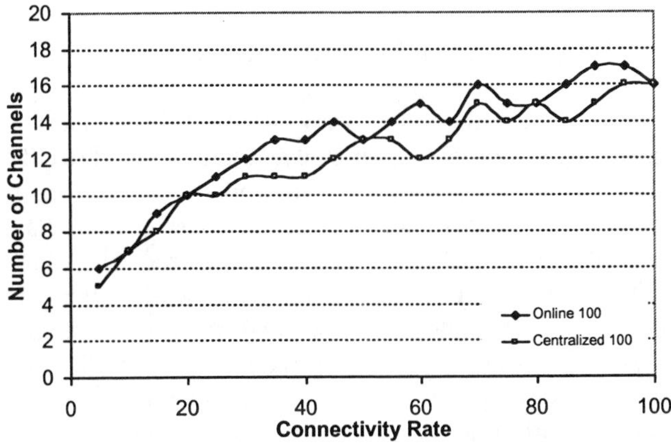

Figure 16: Largest schedule length between the centralized and online algorithm as the connectivity rate changes in a small network containing 3-8 neighbors per sensor.

In the next set of experiments, we evaluate the performance for increasing connectivity. We compare the schedule lengths for small networks (with 100 sensors) and large (with 4900 sensors) networks against the connectivity rate. Like previous experiment, the topology contains 3-8 neighbors. As expected, the schedule length in Figure 17 is larger than the schedules in Figure 16. As the number of subscribers increases, so will the conflicts in trying to use the same channels. Similarly, an increase in the number of neighbors will also increase the schedule length because of the conflicts.

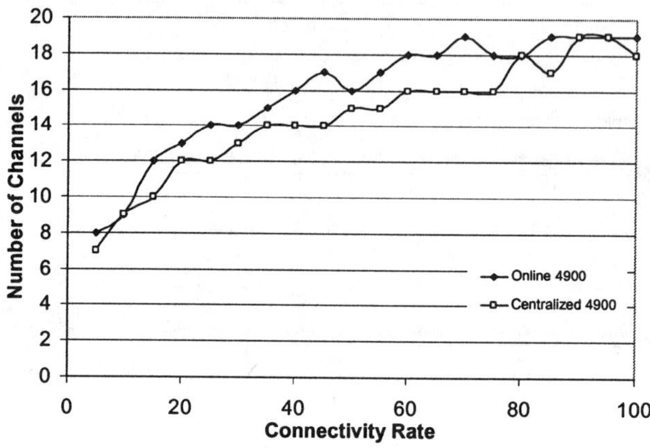

Figure 17: Largest schedule length between the centralized and online algorithm as the connectivity rate changes in a large network containing 3-8 neighbors per sensor.

5. RELATED WORK

In addition to the directly related work discussed in the introduction, there are a number of studies relevant to our study. Liu and Lloyd [LUI01] proposed a distributed online algorithm for ad hoc networks. [SWU00] proposed MAC protocols for ad-hoc networks for a fixed number of channels. [KAN01] proposed distributed and [DAM99] proposed centralized scheduling algorithms for infrastructure based wireless networks. Our work differs from all this previous work as it concentrates on sensor network communication where the data transmissions are well defined using publish/subscribe information. Our algorithm assumes no prior knowledge of the topology or the subscription, and for this reason we depend on the application layer information on publications and subscriptions in order to reduce the number of channels as well as energy consumption by a sensor. In their environment, the data transmission was not well defined like our publish/subscribe topology, which gives us the opportunity to optimize the transmission based on publish/subscribe information coming from the application layer.

6. CONCLUSIONS AND FUTURE WORK

In this paper, we described an online scheduling algorithm and discussed the issues that arise with publish/subscribe sensor network communications, our online algorithm uses message exchanges between sensors to discover its neighbors, to advertise its topic, and to subscribe to interested topics. We generate a directed graph to describe the publish/subscribe relationship of sensors in the network. The sensors then reserve channels using a reservation protocol. Unlike previously proposed offline algorithms, our online algorithm does not assume any initial knowledge of the network topology. Our algorithm provided a performance that is very close to that of the centralized algorithm in terms of schedule length. More importantly, it provided an order of magnitude smaller EICU that indicates efficient use of the channels, which is a major issue in an energy-limited environment. For future work, we will investigate QoS issues, to study in particular how delay and throughput service level agreements can be incorporated into the scheduler.

7. REFERENCES

[CAR01] A. Carzaniga, D.S. Rosenblum, and A.L. Wolf, "Design and Evaluation of a Wide-Area Event Notification Service," *ACM Transactions on Computer Systems*, August 2001.

[WEB03] http://www.davisnet.com

[RAM92] S. Ramanathan and E.L. Lloyd, "Scheduling Algorithms for Multi-hop Radio Networks," *SIGCOMM*, 1992.

[RAM99] S. Ramanathan, "A Unified Framework and Algorithm for Channel Assignment in Wireless Networks," *Wireless Networks*, 1999.

[LUI01] L. Lui, E.L. Lloyd, "A Distributed Protocol for Adaptive Link Scheduling in Ad-hoc Networks," *International Conference on Wireless and Optical Communications,* 2001.

[JJU99] J. Ju and V.O.K. Li, "TDMA Scheduling Design of Multihop Packet Radio Networks Based on Latin Squares," *INFOCOM*, 1999.

[BJO03] R. Bjorklund, P. Varbrand, and D. Yuan, "Resource Optimization of Spatial TDMA in Ad Hoc Radio Networks: A Column Generation Approach," *INFOCOM*, 2003.

[GRO00] J. Gronkvist, "Assignment Methods for Spatial Reuse TDMA," *MOBICOM*, 2000.

[DAM99] S. Damodaran and K.M. Sivalingam, "Scheduling in Wireless Networks with Multiple Transmission Channels," *International Conference on Network Protocols*, 1999.

[SWU00] S. Wu, C. Lin, Y. Tseng, and J. Sheu, "A New Multi-Channel MAC Protocol with On-Demand Channel Assignment for Multi-Hop Mobile ad hoc Networks," *Parallel Architectures, Algorithms, and Networks*, 2000.

[KAN01] V. Kanodia, C. Li, A. Sabharwal, B. Sadeghi, and E. Knightly, "Distributed Multi-Hop Scheduling and Medium Access with Delay and Throughput Constraints," *MOBICOM*, 2001.

Geospatial Applications of Sensor Networks

In-Situ Sensorweb Prototype Demonstrations for Integrated Earth Sensing Applications

P.M. Teillet, A. Chichagov, G. Fedosejevs, R.P. Gauthier, A. Deschamps, T.J. Pultz, G. Ainsley, M. Maloley, and F. Simard

Canada Centre for Remote Sensing
588 Booth Street, Ottawa, Ontario, Canada K1A 0Y7

ABSTRACT

This paper describes initial steps toward building an integrated earth sensing capability that encompasses both remote and in-situ sensing. Initial work has focused on the demonstration of in-situ sensorweb prototypes in autonomous remote operation in the context of monitoring applications in Earth and environmental science. The paper discusses integrated Earth sensing and sensorwebs, and reports on an in-situ sensorweb prototype demonstration in support of flood hazard monitoring in Manitoba, as well as on development plans for a more advanced, heterogeneous sensorweb in support of drought severity and rangeland/crop vigour monitoring in Alberta.

1. INTRODUCTION: INTEGRATED EARTH SENSING AND SENSORWEBS

Our monitoring requirements and responsibilities as nations and as members of the global community continue to multiply. We have some powerful science and technology tools at our disposal but it is not clear that we are using them effectively to tackle the issues before us. In many countries, government agencies in particular have long traditions of excellence in field data acquisition and more recently space-based observations of the Earth. However, such endeavours have been and largely remain resource-intensive activities. Innovative tools need to be developed to provide the time-critical and cost-effective monitoring of complex and dynamic systems essential to support effective decision-making.

At the 2002 World Summit on Sustainable Development, the point was made that "... space-derived information generally needs to be combined with in-situ measurements and models to obtain a holistic picture of the Earth's environment. ... There is no Sustainable Development without adequate information about the state of the Earth and its environment"[a]. Indeed, the

[a] Josef Aschbacher, European Space Agency (ESA).

confluence of advanced technologies for Earth-based sensorwebs,[b,c,d] Earth science satellite webs [1,2], and the power of the Internet will soon provide a kind of global virtual presence [3] or integrated Earth sensing [4,5].

Unlike other distributed sensor networks, sensors in a sensorweb are "smart" enough to share information with each other, modify their behaviour based on collected data, and only report on and/or act on data, aggregates and/or events of interest to the user. In the in-situ context, a "sensorweb" consists of an autonomous wireless network of smart sensors deployed to monitor and explore environments or, more succinctly, "a macro-instrument for coordinated sensing" [6]. With the capability of providing an ongoing virtual presence in remote locations, many sensorweb uses are being considered in the context of environmental monitoring.

The work reported in this paper has been undertaken within the framework of a threefold effort to: (1) design and deploy sensorwebs for ground-based in-situ data acquisition, (2) develop methods to assimilate in-situ and remote sensing data into models that generate validated information products, and (3) facilitate the accessibility of in-situ sensor data and/or metadata from on-line geospatial data infrastructures. The focus of recent work has been on the first of these. This paper describes initial steps towards building an integrated earth sensing capability and focuses on demonstrations of an in-situ sensorweb prototype in remote operation in the context of Earth and environmental science applications. As a first application, a five-node sensorweb prototype was deployed and operated autonomously in 2002-2003 in the Roseau River Sub-Basin of the Red River Watershed in Manitoba, Canada as part of a flood hazard monitoring project. Soil temperature, soil moisture and standard meteorological measurements were accessed remotely via landline and/or satellite telecommunication from the Integrated Earth Sensing Workstation (IESW) in Ottawa. Independent soil moisture data were acquired from grab

---

[b] Pister, K.S.J., Kahn, J.M., and Boser, B.E. "Smart Dust: Wireless Networks of Millimeter-Scale Sensor Nodes", Highlight Article in *1999 Electronics Research Laboratory Research Summary*, Department of Electrical Engineering and Computer Science, University of California, Berkeley, California, 94720 USA, 1999, 6 pages.

[c] Neil Gross, "The Earth Will Don an Electronic Skin", in "21 Ideas for the 21st Century", *BusinessWeek online*, August 30, 1999.

[d] Wireless sensor networks have been identified recently as one of ten emerging technologies that will change the world. Cf. Roush, W. "Wireless Sensor Networks", *Technology Review (MIT's Magazine of Innovation)*, 106(1): 36-37, 2003.

samples and field-portable sensors on days when Radarsat and Envisat synthetic aperture radar (SAR) image acquisition took place. The in-situ data enabled spatial soil moisture estimates to be made from the remotely sensed data for use in a hydrological model for the applications at hand. Another project is underway for the development of a more advanced, heterogeneous sensorweb in support of drought severity monitoring in Alberta.

2. PROTOTYPE WIRELESS INTELLIGENT SENSORWEB EVALUATION

The Prototype Wireless Intelligent Sensorweb Evaluation (ProWISE) project has targeted the field deployment of a sensorweb with full inter-nodal connectivity and remote access and control. It has also tested remote webcam operations and demonstrated telepresence at remote field sites. These deployments do not yet take advantage of fully miniaturized systems but they utilize commercial-off-the-shelf technology and are taking place in real as opposed to controlled environments. The initial prototype sensorweb test-bed consists of five nodes and a base station/hub. Each node has a compact mast with Adcon[e] sensors recording temperature, relative humidity, downwelling solar radiation, rainfall, wind direction, wind speed, leaf wetness as appropriate, soil temperature, and soil moisture (Figure 1).

Figure 1: Fully instrumented ProWISE sensorweb node deployed at Transport Canada Federal Airport site for a groundwater project near Toronto, Ontario. The top end of the soil moisture probe is visible in the foreground. Behind that to the left is the solar panel and in front of the mast is the case containing the MT-2000 satellite transceiver, HC-12 microprocessor, and batteries.

[e] Adcon, http://www.adcon.at/adcon/english/welcome.htm.

Over time, the instrumentation has been augmented by various telecommunication devices, micro-controllers, and other sensors from various vendors. Different wireless telecommunication strategies have been examined. Access and control are Internet web-enabled, remotely operated, and have been tested from the individual nodes to the IESW in Ottawa as well as from the nodes to the hub and then to the IESW. Smart internodal communication is part of the next phase of the ProWISE development in the drought severity monitoring project.

3. FLOOD HAZARD MONITORING: A PROTOTYPE IN-SITU SENSORWEB DEMONSTRATION

There are significant urban populations and infrastructures that are vulnerable to flooding in the Red River Watershed. Consequently, there are ongoing studies to enhance the flood protection infrastructure at Winnipeg, Manitoba. Integrated Earth sensing activities that encompass both remote and in-situ sensing will contribute to enhanced flood mitigation, flood forecasting and emergency planning along the Red River, where there exists the potential for multi-million dollar flood disasters. The in-situ sensorweb work described in this paper is part of an effort to produce hazard and infrastructure assessments and improve real-time monitoring capabilities that contribute to reducing the impacts and costs of flood disasters, and improve decision-making during flood emergencies [7].

The ProWISE sensorweb was deployed in the Roseau River Sub-Basin of the Red River Watershed in September 2002 and remained there throughout the winter months and the flood season until the end of June 2003. The five nodes and hub were distributed across an extent of approximately 50 km. The sensorweb operated autonomously and provided soil temperature and soil moisture measurements plus standard meteorological parameters remotely via landline and/or satellite to the IESW at the Canada Centre for Remote Sensing (CCRS) in Ottawa, Ontario. C-band Synthetic Aperture Radar (SAR) data were acquired from the Radarsat-1 SAR (HH polarization) and the Envisat Advanced SAR (ASAR) (HH and VV polarizations) instruments in order to estimate and monitor soil moisture over large areas [8,9]. The in-situ data were used to help generate spatial soil moisture estimates from the SAR image data for use in the WATFLOOD hydrological model for flood hazard monitoring. Initially, the data collected by the soil moisture sensors (C-Probes) were transmitted with the other sensor data via the Adcon A733 wireless radio modems to the Adcon A840 data storage unit and gateway situated at the hub in Shevchenko School in Vita, Manitoba. These data were then transmitted via a modem connection by a landline to the IESW. The C-Probe terrestrial wireless link to the A840 was replaced by a Vistar MT-2000 satellite transceiver link for trial periods (Figure 2). The C-Probe communicated with the MT-2000 via a programmed HC-12 microprocessor. This put each soil moisture sensor from each individual node "on the air" in real time and constitutes an important step in demonstrating different modes of sensorweb deployment.

4. DROUGHT SEVERITY MONITORING: TOWARD A HETEROGENEOUS IN-SITU SENSORWEB DEMONSTRATION

Detecting onset of drought and, more generally, predicting and assessing crop growth and rangeland productivity has very important commercial, environmental and community benefits. A new project to test new capabilities for drought severity monitoring in Alberta is focusing on three key components for success: (1) a smart sensorweb that provides autonomous and continuous in-situ measurements and overcomes the spatial-temporal sampling problem; (2) smart vegetation modeling agents that integrate all available relevant information to yield economically and ecologically valuable information; (3) an OpenGIS Consortium (OGC) compliant, Internet-based infrastructure that facilitates communication between sensor nodes and servers and makes the data and services accessible to all who can benefit from them. In this project, the current ProWISE nodes will become the base stations for local-area sensorwebs consisting of many smaller sensor nodes. Thus, hierarchically, there will be at least two levels to the in-situ sensorweb, encompassing different types of sensors and sensor nodes, making it a so-called heterogeneous sensorweb.

5. CONCLUDING REMARKS

Earth science sensorweb data have the potential to become an integral part of government policy and decision support domains. The work reported in this paper has taken some initial steps towards demonstrating new approaches to the time-critical and cost-effective monitoring of complex and dynamic systems. Nevertheless, much more needs to be done to provide a more solid basis for issue-specific decision support, including smarter, smaller and cheaper sensor systems for monitoring, the integration of time-critical in-situ sensor data and/or metadata into on-line geospatial data infrastructures, and the generation of validated data and information products derived from the fusion of in-situ and remote sensing data and their assimilation into models. Ongoing challenges in such endeavours include insufficient resources to put in place capabilities for integrated assessment and the potential future shortfall of highly qualified science and technology personnel.

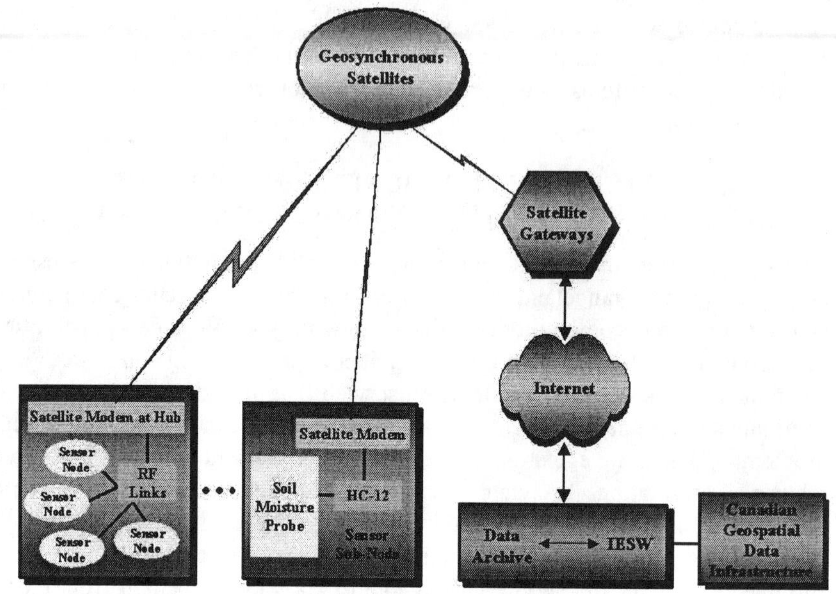

Figure 2: Logical data flow of the deployment configuration and satellite telecommunication routing between sensors in the field and the Integrated Earth Sensing Workstation (IESW) in Ottawa. Except for proof-of-concept trials, the hub uses landlines to communicate with the IESW to save costs.

REFERENCES

[1] NASA. *Exploring Our Home Planet: Earth Science Enterprise Strategic Plan*, NASA Headquarters, Washington, DC, 2000.

[2] Zhou, G, Baysal, O., Kafatos, M., and Yang, R. (Editors). *Real-Time Information Technology for Future Intelligent Earth Observing Satellites*, ISBN: 0-9727940-0-X, Hierophantes Publishing Services, Pottstown, PA, 2003.

[3] Delin, K.A. and Jackson, S.P. "The Sensor Web: A New Instrument Concept", *Proceedings of SPIE's Symposium on Integrated Optics*, San Jose, CA, January 2001. (See also http://sensorwebs.jpl.nasa.gov/ resources/sensorweb-concept.pdf)

[4] Teillet, P.M., Gauthier, R.P., Chichagov, A., and Fedosejevs, G. "Towards Integrated Earth Sensing: Advanced Technologies for In Situ Sensing in the Context of Earth Observation", *Canadian Journal of Remote Sensing*, 28(6): 713-718, 2002.

[5] Teillet, P.M., Gauthier R.P., and Chichagov A. "Towards Integrated Earth Sensing: The Role of In Situ Sensing", Chapter 2, pp. 19-30, in *Real-time Information Technology for Future Intelligent Earth Observing Satellites*, (Eds.). Zhou, G., Baysal, O., Kafatos, M., and Yang, R., ISBN: 0-9727940-0-X, Hierophantes Publishing Services, Pottstown, PA, 2003.

[6] Delin K.A. "The Sensor Web: A Macro-Instrument for Coordinated Sensing." *Sensors,* 2: 270-285, 2002. (See also http://sensorwebs.jpl.nasa.gov/resources/Delin2002.pdf)

[7] Wood, M.D., Henderson, I., Pultz, T.J., Teillet, P.M., Zakrevsky, J.G., Crookshank, N., Cranton, J., and Jeena, A. "Integration of Remote and *In Situ* Data: Prototype Flood Information Management System", *Proc. of the 2002 IEEE Geoscience and Remote Sensing Symposium (IGARSS 2002) and the 24th Canadian Symposium on Remote Sensing*, Toronto, Ontario, Volume III, pp. 1694-1696. (also on CD-ROM, 2002.)

[8] Wigneron, J.-P., Calvet, J.-C., Pellarin, T., Van de Griend, A.A., Berger, M., and Ferrzzoli, P. "Retrieving Near-Surface Soil Moisture From Microwave Radiometric Observations: Current Status and Future Plans", *Remote Sensing of Environment*, 85(4): 489-516, 2003.

[9] Boisvert, J.B., Pultz, T.J., Brown, R.J., and Brisco, B. "Potential of Synthetic Aperture Radar for Large Scale Soil Moisture Monitoring", *Canadian Journal of Remote Sensing*, 22(1): 2-13, 1996.

GeoSWIFT: Open Geospatial Sensing Services for Sensor Web

Vincent Tao, Steve Liang, Arie Croitoru, Zia Moin Haider, and Chris Wang

Geospatial Information and Communication Technology (GeoICT) Lab
Department of Earth and Atmospheric Science, York University
4700 Keele Street, Toronto, ON, Canada M3J 1P3
E-mail: {tao, liang, arie, ziamoin, chunwang}@yorku.ca
URL: www.geoict.net

1. INTRODUCTION

With the presence of cheaper, miniature and smart sensors; abundant fast and ubiquitous computing devices; wireless and mobile communication networks; and autonomous and intelligent software agents, the Sensor Web has become a clear technological trend in geospatial data collection, fusion and distribution. The Sensor Web is a Web-centric, open, interconnected, intelligent and dynamic network of sensors that presents a new vision for how we deploy sensors, collect data, and fuse and distribute information. The Sensor Web is an evolving concept with many different research efforts working to define the possibilities.

The Sensor Web offers full-dimensional, full-scale and full-phase sensing and monitoring of Earth at all levels: global, regional and local. The Sensor Web is a revolutionary concept toward achieving collaborative, coherent, consistent and consolidated sensor data collection, fusion and distribution. Such sensors include flood gauges, air-pollution monitors, stress gauges on bridges, mobile heart monitors, Webcams, and satellite-borne earth imaging devices. The Web is considered a "central computer" that connects enormous computing resources. The Sensor Web can similarly be thought of as a global sensor that connects all sensors or sensor databases.

2. CHARACTERISTICS OF SENSOR WEB

The Sensor Web is an evolving concept with many different research efforts working to define the possibilities. Examples of pioneering work include sensor pods (Delin and Jackson 2000), Smart Dust (Warneke, Atwood et al. 2001) and integrated earth sensing (Teillet, Gauthier et al. 2002).

Inspired by significant advances in sensors, communication, computing and positioning technologies, the Sensor Web is being developed by connecting heterogeneous sensors or many proprietary sensor networks. To succeed, the Sensor Web must be interoperable, intelligent, dynamic, and scalable.

Interoperable

The Sensor Web is achieved by connecting the distributed, dynamic and heterogeneous in-situ and remote sensors to an open, interconnected network - the Web. The Sensor Web is a universe of network-accessible sensors, sensory data and information. Just like building a Web system, developing the Sensor Web requires thorough and careful design of system hierarchy, registry, Internet Protocol domain services and applications. Implementation of interoperability requires commonly accepted standards and specifications, and the Open GIS Consortium http://www.opengis.org has pioneered much of this activity.

Intelligent

Such intelligence comes from sensor connectivity, just like human intelligence comes from connected neurons in the brain. Sensor networks forage for information the way ants forage for food. By linking existing databases or previously sensed data with the Sensor Web, we will dramatically increase the efficiency and performance of sensing, monitoring and change detection.

Dynamic

Thanks to wireless communication and real-time positioning technologies, sensors no longer need to be fixed at a certain location. By removing physical constraints, sensors can be mobile, placed anywhere or even be seeded for large-area monitoring. As sensors are position aware, their location and movement can be tracked continuously via wireless networks.

Scalable

The Sensor Web is intended to offer massive sensing by interconnecting a large array of sensors. Therefore, the Sensor Web must not be inherently restricted to contain a specific number of nodes or operate within a predefined area (Sohrabi, Merrill et al. 2002).

Mobile

Location is an essential component of the Sensor Web. When and where sensor data is observed is of equal value to the sensor data itself. There are wide options to integrate positioning technologies (e.g., GPS, A-GPS, Internet GPS, radio-frequency identification, real-time locating system, cellular network positioning, etc.) with sensor networks. There are important research topics in location-based routing of sensors as well as optimized sensor network topology and configuration.

3. OPEN SENSOR WEB SERVICES

There are many technical issues involved in building the Sensor Web framework. From a system architecture viewpoint, the Sensor Web involves the following layers:

- Sensor layer-sensor design, materials, miniaturization, energy consumption, etc.
- Communication layer-networking, protocol, topology, etc.
- Information layer-agents, management, fusion, distribution, etc.

Depending on the properties of sensors, geographic coverage, network access capabilities and, more importantly, domain applications, the physical architecture (i.e., the first three layers) can be very different. The information layer serves as a backbone and shares a commonality. This layer is a gateway to integrate and fuse observations from spatially referenced sensors. It connects widely distributed in-situ sensors and remote sensors over wired or wireless networks. Interoperability becomes a key to enable the information layer's integration capability.

4. WEB SERVICE ARCHITECTURES

We use a new technology framework, Web Services, to design the architecture of our open geospatial sensing service for Sensor Web: GeoSWIFT. Web Services represent the convergence between the service-oriented architecture (SOA) and the Web (W3C, 2002). SOA has evolved over the last 10 years to support high performance, scalability, reliability, and availability. However, traditional SOA are tightly coupled with specific protocols. Each of the protocols is constrained by dependencies on vendor implementations, platforms, languages, or data encoding schemes that severely limit interoperability. The Web Services architecture takes the advantageous features of the SOA and combines it with the Web. The Web supports universal communication using loosely coupled connections. Web protocols are completely vendor-, platform-, and language-independent (Systinet Corp., 2002). Web Services support Web-based access, easy integration, and service reusability. Web Services satisfies our requirements to build a geospatial infrastructure for Sensor Webs with openness, interoperability, and extensibility. Web Services now are developing as standards in W3C group and will become future standards for the Web.

5. OGC WEB SERVICE

OGC Web Services are an evolutionary, standards-based framework that will enable seamless integration of a variety of online geoprocessing and location services. OGC Web Services will allow distributed geoprocessing systems to communicate with each other using technologies such as XML and HTTP. This means that systems capable of working with XML and HTTP will be able to both advertise and use OGC Web Services.

OGC has taken a leadership role in developing and building a unique and revolutionary open platform for exploiting sensors. OGC's goal is to make all

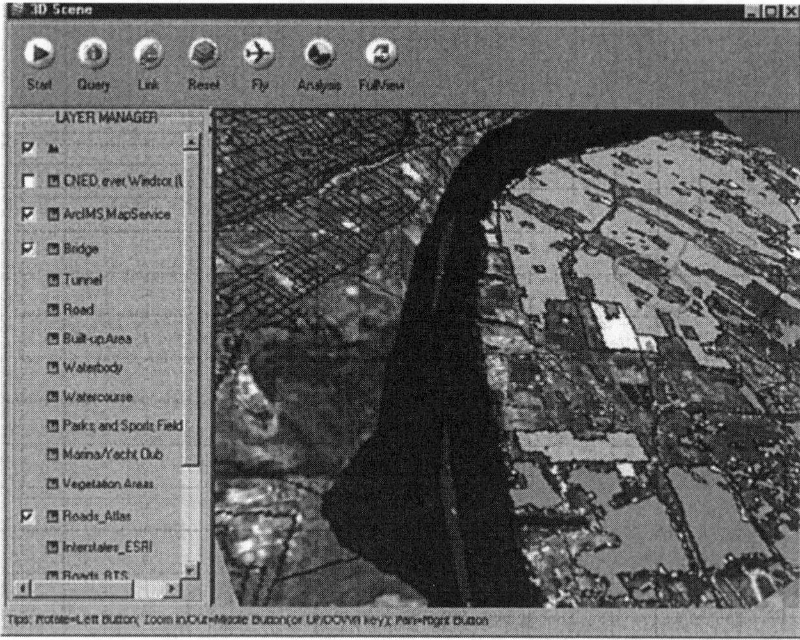

Figure 1: GeoServNet OGC WMS 3D Viewer.

types of sensors, instruments and imaging devices as well as repositories of sensor data discoverable, accessible and, where applicable, controllable via the Web. OGC Web Services will allow future applications to be assembled from multiple, network-enabled geoprocessing and location services. This capability will be possible because rules will be established for these services to advertise the functionality they provide and how to send service requests via open, standard methods. In this manner, OGC Web Services provide a vendor-neutral interoperable framework for web-based discovery, access, integration, analysis, exploitation and visualization of multiple online geodata sources, sensor-derived information, and geoprocessing capabilities. As a

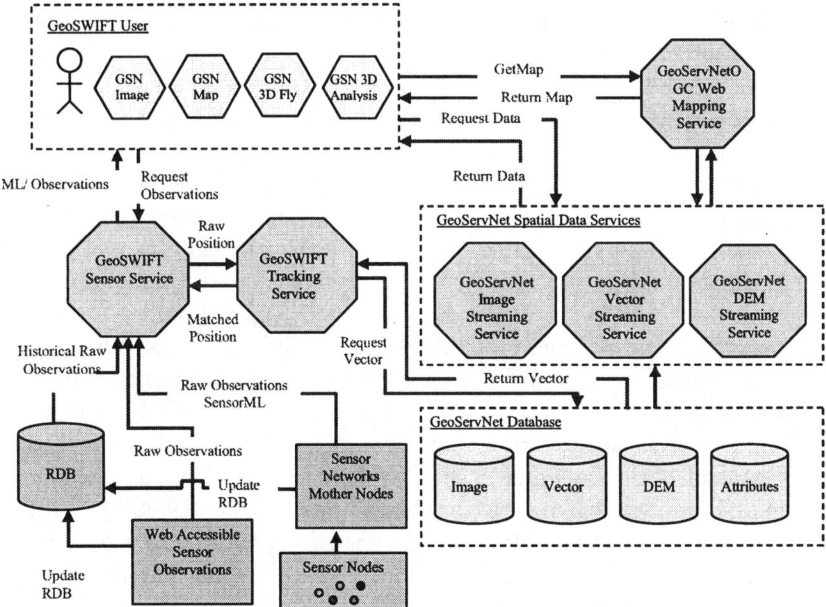

Figure 2: System Architecture of GeoSWIFT.

member of OGC, GeoICT Lab is actively involved in the development and implementation of OGC specifications. Figure 1 shows a screen capture of the GeoServNet OGC WMS 3D Viewer developed at GeoICT lab for OGC CIPI (Critical Infrastructure Protection Initiative). GeoSWIFT is based on the same technology, GeoServNet, for the components of server and 2D/3D sensor viewer.

6. GEOSWIFT

GeoSWIFT is an ongoing project sponsored by GEOIDE, PRECARN and CRESTech. The goal of GeoSWIFT is to build a geo-spatial infrastructure to connect distributed sensor networks for the sharing, access, exploitation, and analysis of sensing information. The development of GeoSWIFT involves both software engineering and communication development work. Various intelligent sensors will be deployed to the fields and several different real-world applications of the system, such as snow depth monitoring, crop yielding prediction, nuclear safety, meteorological services, and real-time tracking, will be evaluated with our industrial and government partners, such as MacDonald Dettwiler Inc., Canadian Centre of Remote Sensing (CCRS), and Canadian Nuclear Safety Commission (CNSC).

GeoSWIFT is built upon GeoServNet, a GIServices system developed at GeoICT lab at York University. GIService allows users to access, assemble, or even "rent" geoprocessing components that are distributed across a network via standard web browsers (Tao 2001). Figure 2 shows the system architecture of GeoSWIFT system.

The octagons in Figure 2 are service providers. GeoSWIFT Sensor Service is the portal of sensors to provide web services interfaces for sensor networks, its observations and the related geo-spatial information as well. Depending on the properties of sensors, geographic coverage, network access capabilities and, more importantly, domain applications, the physical architecture of sensor networks can be very different. GeoSWIFT Sensor Service serves as a wrapper which hides the different communication protocols, data formats and standards of sensor systems behind this layer and provides a standard interface for clients to collect and access sensor observations and manipulate them in different ways. Various clients only need to follow the OGC interfaces which are provided by GeoSWIFT Sensor Server, and do not need to deal with the annoying protocols of various sensor networks (Liang, Tao et al. 2003). Table 1 shows the four interfaces defined by OGC (OGC 2003).

The hexagons in Figure 2 are viewers for Sensor Web in the architecture of GeoSWIFT. GeoSWIFT Viewer is built based on GeoServNet Viewer. GeoServNet Viewer is a 2-D/3-D Java Web GIService viewer, and designed for streaming large amount of spatial data via Internet.(Han, Tao et al. 2003) Sensor Web comprises a large number and different type of in-situ and remote sensors which will produce massive and diverse data that can challenge our ability to process and manage the data. How to accommodate various sensor types in a viewer and deliver valuable information to the end user becomes a challenging task as well. For example, a 3D viewer is needed for the visualization a series of deployed balloons of differing buoyancy to determine air movement. The multi-dimension and diversity of sensor platforms and its observations, however, offers the opportunity to take advantage of interactive 3D visualization techniques that can improve the efficiency and accuracy of processing, and provide an unprecedented perspective of sensor observations

7. CONCLUSIONS

The Sensor Web presents endless opportunities for adding a sensory dimension to the Web's globe-encircling virtual nervous system. This has extraordinary significance for science, environmental monitoring, transportation management, public safety, homeland security, defense, disaster management, health and many other domains of activity.

The fundamental revolution in the Sensor Web vision lies in its interoperable, intelligent, dynamic, flexible, and scalable connectivity. It is an evolving framework enabled by many emerging technologies. Our understanding of the Sensor Web is in its infancy. GeoSWIFT is just a beginning, and we have already seen many exciting applications of it. Just like the Web 10 years ago, and it is only a matter of time before the Sensor Web joins the fabric of our lives and becomes indistinguishable.

Table 1: Interfaces Provided by GeoSWIFT Sensor Server

Requests	Responses
GetCapabilities	The responding XML of service's capabilities conforms OGC Service Information Model Schema (OGC 2003), provides detailed information for a client to access the service. The provided information includes Service Type, Service Instance, Content Type, and Content Instance.
GetObservations	The responding XML of GetObservation request is encoded conforming to GML and O&M schema. It contains values, units, and locations of the requesting sensor observations.
DescribePlatform	The XML response describes the sensor platform, and conforms to SensorML schema. An example of a sensor platform can be a plane tht carries a camera, several inertial sensors, and meteorological sensors. The plane is the platform.
DescribeSensor	The XML response contains detailed information of sensor characteristics encoded in SensorML. The sensor characteristics can include lists and definitions of observables supported by the sensor.

8. REFERENCES

Delin, K.A. and S.P. Jackson, 2000. Sensor Web for In Situ Exploration of Gaseous Biosignatures. *2000 IEEE Aerospace Conference*, Big Sky, MO.

Han, H., V. Tao et al., 2003. Progressive Vector Data Transmission. *6th AGILE Conference on Geographic Information Science*, Lyon, France, Presses Polytechniques et Universitaires Romandes.

Liang, S.H.L., V. Tao et al., 2003. The Design and Prototype of a Distributed Geospatial Infrastructure for Smart Sensor Webs. *6th AGILE Conference on Geographic Information Science*, Lyon, France, Presses Polytechniques et Universitaires Romandes.

OGC, 2003. *OWS 1.2* Service Information Model, *OpenGIS Interoperability Program Report.* J. Lieberman.

OGC, 2003. *Sensor Collection Service.* T. McCarty.

Sohrabi, K., W. Merrill et al., 2002. Scaleable Self-Assembly for Ad Hoc Wireless Sensor Networks. *IEEE CAS Workshop on Wireless Communications and Networking*, Pasadena, CA.

Tao, V. 2001. Online GIServices. *Journal of Geospatial Engineering*, 3(2): 135-143.

Teillet, P.M., R.P. Gauthier et al., 2002. Towards Integrated Earth Sensing: Advanced Technologies for In Situ Sensing in the Context of Earth Observation. *The Canadian Journal of Remote Sensing,* 28(6): 713-718.

Warneke, B., B. Atwood, et al., 2001. Smart Dust Mote Forerunners. *Proc. 14th IEEE International Conference in Micro Electro Mechanical Systems (MEMS 2001),* Interlaken, Switzerland.

Symbiote: An Autonomous Sensor for Urban Operation Imagery

Benoit Ricard[1], Maj Michel Gareau[1], and Martin Labrie[2]

[1] Defence Research and Development Canada –Valcartier.
[2] Département de génie électrique et informatique, Université Laval, Canada.

ABSTRACT

During their missions, the Canadian Forces (CF) routinely performs different types of mounted patrols (reconnaissance, social or road scouting). Furthermore, the trends show that armed forces will be more and more required to carry out missions in complex environments such as urban zones. Today's technologies allow specialized sensors to gather large amounts of data, which, when adequately organized could facilitate CF mission planning and safe execution of them. Deployed troops depend directly on HUMINT (Human Intelligence) to collect data to feed Intelligence cells. Unfortunately, this kind of information lacks precision and is difficult to merge into operational databases. The goal of this project is to develop a semi-autonomous sensor to support intelligence cell data gathering.

1. INTRODUCTION AND OVERVIEW OF THE SYSTEM

In order to provide the intelligence cells with information, reconnaissance troops should record their observations and take imagery during patrols. The pace of the intelligence cycle depends directly on the size of the data gathering, processing and analysis effort and on the manpower in the field. To increase the pace and relieve the soldier from repetitious tasks requiring precision and concentration, we designed a novel sensor called Symbiote. The latter is composed of a sensor head small enough to be concealed on a patrolling vehicle and a mission planning and management (MPM) software package on a remote control station.

The operation of the Symbiote System is phased over a 3-step process: Mission planning, data acquisition and data exploitation. Mission planning is done on the control station with a graphical software interface that circumscribes and designates objectives of interest (building, bridge or check point) on a topographical map or aerial photo. Once completed, the mission plan is then uploaded from the control station to the sensor head via an Ethernet RF link. At this point, the data acquisition phase begins and the sensor head autonomously collects data on the target specified in the mission plan. As the platform is moving, the target list is continuously sorted and prioritized according to the platform position, target range and view-angle criteria. Line-of-sight is detected and confirmed by comparing range stored in the mission plan with those provided by a laser rangefinder. When the mission is accomplished, imagery and contextual data that were collected is retrieved via the RF link, pre-processed and integrated in the operational database in the form of metadata. Detailed and precise information about the selected objective is available to the MPM software for data exploitation. For instance, partial imagery can be stitched together and ultimately extract 3D models of objects from meta-data collected by Symbiote.

This document reports the first stage of the project in which we have developed a prototype of the camera head and the algorithms to drive the autonomous imaging of targets. The simulator developed for validating the Symbiote system is also described in this paper.

2. FUNCTIONAL DESCRIPTION OF THE SYSTEM

2.1. Mission Planning and Management Software

The first step in using the system is to establish an information-gathering plan for the intelligence cells, using the mission planning and management (MPM) software. All troop movements are traditionally planned jointly by the operations and intelligence cells of the formation being deployed. During this planning stage, the critical elements and information collection tasks are identified and defined. The MPM software is used to graphically define the sensor's data collection tasks. Stationary targets (points, areas or volumes) will be defined by their geographical coordinates to enable the sensor, as the mission proceeds, to determine the right moment for acquiring images as a function of position, direction and speed of the host vehicle.

Building a target database

When the program is launched, a new mission is created, generating a blank database that will contain the definition of the various targets to be imaged, references to the mapping media used in preparing the data collection plan and, later on, the images collected and their contextual data. As mentioned above, the program allows us to graphically define a volume encompassing a fixed structure such as a building, bridge or other structure represented on a map or a photo. This volume is constructed using a succession of lines delimiting the target. Once an area is delimited by the lines, the vertical projection required to create a volume is entered into the program. This volume is the information which is transmitted to the sensor and enables it to plan its images acquisition. When a target is defined, a record identified by the target's name is created in a database. Consider, for example, a building as the target. Each segment delimiting the target represents a wall of the building. The segment is the basic unit of the target definition system and information is attached to each segment. In the creation of targets using the program, a great deal of information, such as the length of walls and the various vectors used in imaging, is pre-calculated to speed up the work during the mission. The sensor module subsequently uses this information to autonomously gather data.

2.2. The Symbiote Sensor and Data Acquisition

The sensor module, named Symbiote, will be installed on a patrol vehicle, but will be completely independent of the vehicle resources. As the sensor's name suggests, it will exist in symbiotic association with the vehicle carrying it; the vehicle is its means of locomotion. Communication between the sensor and the outside world will be provided by an RF link. Once the mission has started, the sensor is self-sufficient. Thus the sensor can autonomously collate various types of digital data—distance measurements, sounds and images of urban points of interest previously identified and preloaded into the sensor prior to the patrol. However, an operator can suspend the programmed mission and change to manual mode to collect unprogrammed data (Figure 1).

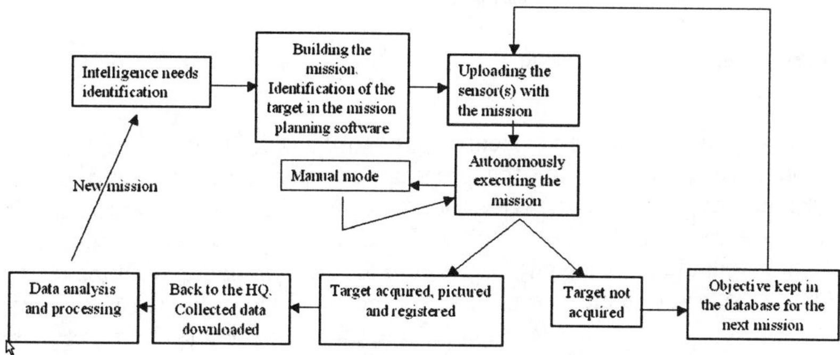

Figure 1: Flowchart of the data gathering process from the planning to analysis.

Operation of the Sensor

For a better understanding of the complexity of the imaging process and of the elements which the camera module must contain, it is important at this point to explain the operating environment of the system. To perform its task properly, the camera must know its position and absolute orientation (in relation to the ground) at all times during its mission. The required positioning parameters, provided by a GPS unit, are latitude and longitude, as well as the error on the position obtained. Unless the terrain is very hilly or the target is very far away, the elevation is less important and will not be taken into account in the early phases of the project. Determining the orientation of the camera is a trickier task, as the vehicle is always moving. For the bearing, an electronic compass is appropriate and sufficient as long as the errors inherent in its use are taken into consideration in the calculations. For the pitch angle of the camera, we must use a combination of gyroscopes, clinometers and the compass to obtain an absolute value for orientation. The roll (rotation of the image around the optical axis) is less important and can be corrected afterward. In addition, the gyroscopes can be used to control camera attitude (to compensate for vehicle movement) during imaging (Figure 2).

- Colour camera, motorized iris and focus
- Laser rangefinder
- GPS positioning
- Ethernet RF communications
- Heading by electronic compass + 2-axis tilt
- Autonomous operation for 4 to 6 hours
- Directional microphone*
- 2-axis gyroscopically stabilized mounting*
- Manual control station on a pocket computer*
- Can be installed on all vehicles using magnetic feet or clamps.

*In phase 2 of the project

Figure 2: Concept of sensor head.

Once we have the position and orientation parameters for the camera, we must determine whether we can see the programmed target. For example, during mission planning, we have marked on a map or aerial photo the perimeter of a building which we would like to image. In performing its mission, the sensor may be at a suitable distance and position to acquire an image, but blocked by a building or other obstacle. We must therefore ensure that the target is visible for imaging. A simple, effective way of doing this is to use a laser rangefinder. By comparing the target-vehicle distance as determined by the map and GPS with the one returned by the rangefinder, we can ascertain whether we are in a position to acquire the image. Furthermore, this information will be useful during the 3D reconstruction phase to lift the scale ambiguity inherent in the chosen technique.

Imaging Criteria and Parameters

Once the targets have been identified, the mission is uploaded into the sensor module onboard the vehicle and the mission is started. From this moment on, the sensor's onboard computer retrieves all the segments (walls) comprising the targets and orders them according to two criteria: distance to the target and camera angle relative with the normal to the wall. A third criterion comes into play during imaging: whether the view of the segment to be imaged is unobstructed. If a segment is blocked from view during the imaging attempt, it is placed on a waiting list and regularly revisited to determine whether an unobstructed view is available.

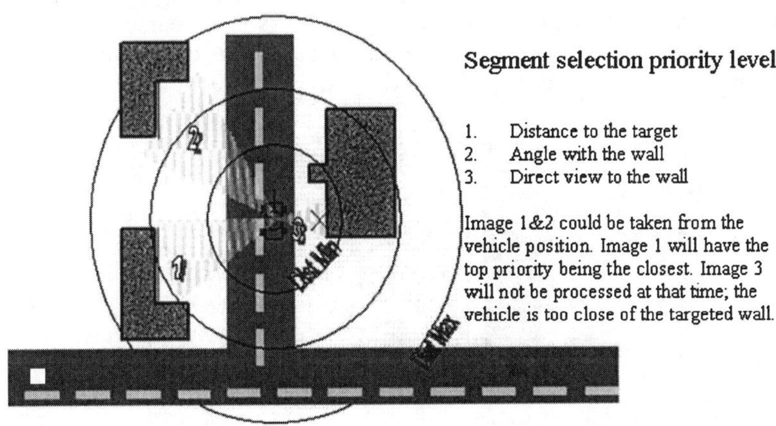

Figure 3: Selection of target according to priority level.

To enable the 3D reconstruction of targets, it was decided that each segment would be imaged from three different angles: 45°, 0° and -45° (angle between the optical axis of the camera and the normal to the segment to be imaged). For this application, it was also decided that three sequences of images would be used to reduce the probability of obstruction by objects, thereby providing an alternative in this case. Also, these image sequences taken around those three angles will be used to track features on wall segments and extract 3D information. In addition, the minimum distance at which a segment is imaged was set at five times the GPS position error (CEP). For example, if the position error is 5 m, no image will be taken at less than 25 m from a target (Figure 3).

Furthermore certain segments may only be partially captured by a single image, so that several images must be obtained and tied together in a mosaic to represent the whole wall (Figure 4). New imaging control parameters must therefore be introduced, such as the camera field of view and the degree of image overlap versus the GPS CEP.

Processing the data

At the end of the mission, the information collected is retrieved from the sensor (through an RF link) and pre-processed automatically to facilitate and speed its integration into the mission databank. Using the sensor's data, highly detailed information can be retrieved concerning the intelligence-gathering targets (buildings, bridges, control points, etc.). The MPM software algorithms ensures the processing of the data to construct images of a broad target based on a mosaic of small images or, using stereo from motion techniques, to build a 3D model of a building that is correctly positioned geographically. Other data, such as sounds or unique images, will also have their content enriched by the contextual data from the sensor (distance to the target, time, lat/long position, etc.). This aggregate information is then referred to as metadata.

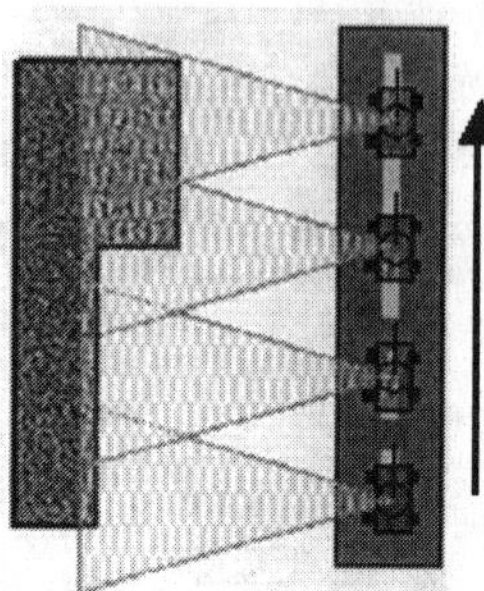

The imaging process should take into account the field of view of the camera, the distance to the wall segment, the positioning error, the optimal overlap between images and the angle to the wall normal.

Figure 4: Multi-images wall segment.

3. BUILDING A SYSTEM PROTOTYPE

After developing the autonomous sensor concept, we focused on its practical implementation: how quickly, with limited financial and human resources, could we validate the concept and develop the basic Symbiote elements? The obvious answer to the first question was simulation, while for the second the solution was to produce a prototype using components available in the laboratory.

3.1. Mission Planning and Simulation Software

The software package was developed in three versions: a server version for mission planning, a client version, similar but running on the sensor platform for debugging purposes and an embedded client version, without any graphic interface. The server version of the software has three functions: mission planning and (sensors) management, data analysis, and imaging simulation. The simulator allowed us to validate the various imaging strategies by simulating the movement of a vehicle and the field of view on a digital map. Simulation consisted of selecting targets on the map for imaging, drawing a patrol route for the vehicle and launching the simulation process. The simulator runs in real time: the movements of the vehicle, the camera head and the imaging process depend on physical parameters fed into the model. Among other things, rotation of the camera head must be anticipated to aim the sensor just a few seconds before taking an image and not continuously, in order to save energy (and avoid interfering with the compass). To accomplish this, we must calculate the time needed to reorient the sensor head in time to aim at the target for imaging. The point from which the next image will be taken is continuously calculated as a function of vehicle speed and heading. If the vehicle changes direction, the point moves and the camera is reoriented.

Figure 5 shows a section of the simulator screen recorded during a mission. The graphic elements represented are the vehicle path, the camera field of view, the calculated point for taking the next image and coloured lines to represent the imaged portion of the wall segments. The simulator was used to successfully develop and validate the various imaging algorithms and strategies. However, practical implementation of the system under actual conditions of use is required to validate the performance achieved by simulation.

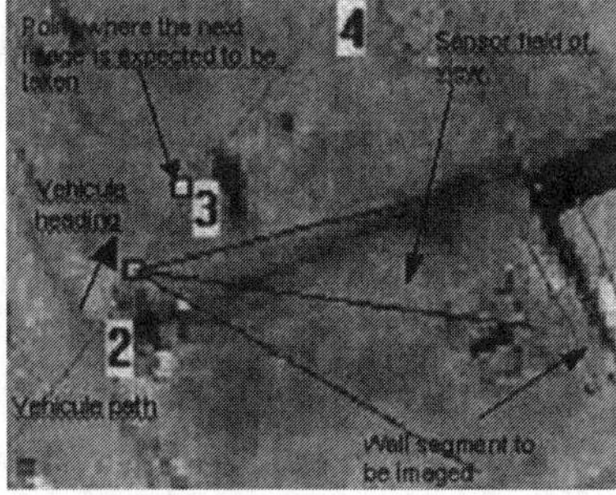

Figure 5: Autonomous data gathering simulation and rehearsal.

3.2. Building the Sensor Head

Using simple hardware that was commercially available and/or already in the laboratory, we produced an initial operational prototype of the Symbiote sensor head in a few weeks (Figure 6). The elements of the system are as follows:

- Matrox 4Sight II PIII, 650 MHz imaging computer
- Leica Vector rangefinder, 1550 nm laser, RS-232 communication
- Basler A302fc CCD camera, colour (Bayers filter), 756 x 512 pixels
- Advanced Orientation Systems EZ-Compass-3 electronic compass.
- Garmin PC-104 GPS (model GPS25 LVS)

All elements were installed on the sensor mount of a mobile robot. The sensor mount allows the simple attachment and orientation of several sensors on the mobile platform. Communication with the platform is by an Ethernet RF link (IEEE 802.11b). For experimental purposes, the robot represents the reconnaissance vehicle, while the sensor mount, rangefinder and CCD camera represent the Symbiote sensor head. An interesting point is that we added a module to the mission planning software to generate the robot's path; thus the robot receives its reconnaissance mission at the same time as the Symbiote sensor.

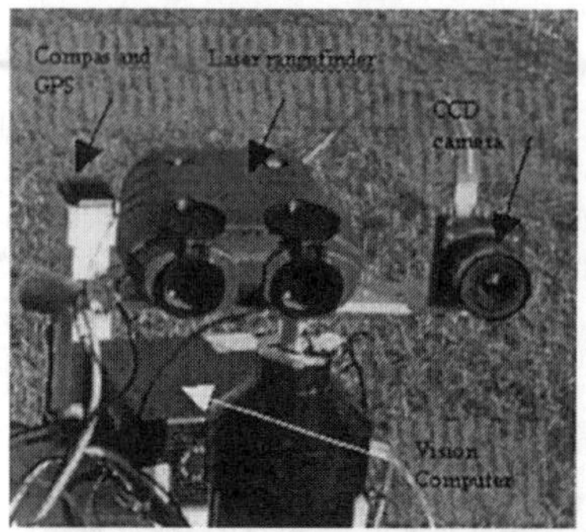

Figure 6: Symbiote prototype demonstrator. Left: sensor head. Right: remotely operated platform.

4. CONCLUDING REMARKS

During the first phase of this project, our aim was to establish the elements of the system and begin testing the various concepts forming Symbiote. Using the simulator, we have already been able to validate the imaging strategies relative to the various parameters and limitations introduced. Simulation of the system led us to the discovery and correction of certain deficiencies it contained and to validation of the photographic coverage of an area of operation. The simulator constructed involves a rudimentary model of the environment; it was not possible, among other things, to simulate obstructions caused by objects, for the simple reason that the only objects introduced were the buildings to be imaged.

Our first outdoor tests revealed the limitations of the robotic platform. We are now installing the camera and laser rangefinder on a pan and tilt unit that will in turn be installed on a vehicle for new outdoor tests. Numerous points need to be addressed in the subsequent phases of the project, such as manual control of imaging, whether the system should be monobloc or in two units, etc. Currently, wall segments are imaged from three fixed angles. It must be determined whether this is necessary and sufficient for the needs of operations, or whether we should provide for an arbitrary number of images that can be captured from arbitrary angles. Finally, the current metadata are specific to our system and appropriate for our application. However, we must improve the current dataset to supply and interface with the command and control systems developed and implemented for the Canadian Forces and Allies.

Author Index

Index

E

H

I

J

K

L

M

Q

R